Ceramic Armor and Armor Systems II

Ceramic Armor and Armor Systems II

Ceramic Transactions Volume 178

Proceedings of the 107th Annual Meeting of The American Ceramic Society, Baltimore, Maryland, USA (2005)

Editor
Eugene Medvedovski

Published by
The American Ceramic Society
735 Ceramic Place, Suite 100
Westerville, Ohio 43081
www.ceramics.org

Ceramic Armor and Armor Systems II

For information on ordering titles published by The American Ceramic Society, or to request a publications catalog, please call 614-794-5890, or visit www.ceramics.org

Printed in the United States of America.

ISBN 1-57498-248-6

10 09 08 07 06 5 4 3 2 1

Contents

Preface vii

Ceramic Armor Development and Study

Advanced Ceramics for Personnel Armor: Current Status and Future 3
Eugene Medvedovski

Characterization of AlON™ Optical Ceramic 19
Thomas M. Hartnett, Charles T. Warner, Donald Fisher, and Wayne Sunne

Synthesis of Dense B_4C-SiC-TiB_2 Composites 37
S. Hayun, N. Frage, H. Dilman, V. Tourbabin, and M.P. Dariel

Biomorphic Reaction Bonded Silicon Carbide Ceramics for Armor Applications 45
Bernhard Heidenreich, Michaela Gahr, and Eugene Medvedovski

Reaction-Bonded SiC Composites without Residual Si 55
John P. Hurley, Versha Singh, and John A. Hamling

Microstructural Engineering of the Si-C-Al-O-N System 63
R. Marc Flinders, Darin Ray, Angela Anderson, and Raymond A. Cutler

Ceramic-Polymer Composites for Ballistic Protection 79
Paolo Colombo, Eugene Medvedovski, and Francesco Zordan

Grain Boundary and Triple Junction Chemistry of Silicon Carbide with Aluminum or Aluminum Nitride Additive 91
Edgardo Pabit, Samantha Crane, Kerry Seiben, Darryl P. Butt, Darin Ray, R. Marc Flinders, and Raymond A. Cutler

Defect Engineering of Samples for Non-Destructive Evaluation (NDE) Ultrasound Testing 103
Raymond Brennan, Richard Haber, Dale Niesz, and James McCauley

Defining Microstructural Tolerance Limits of Defects for SiC Armor 109
Memduh Volkan Demirbas and Richard A. Haber

Fracture Mechanism of Ceramic Armor

Characterization of Subsurface Damages of Static and Dynamic Indented Armor Ceramics: SiC and Si_3N_4 123
Jong Ho Kim, Young Gu Kim, and Do Kyung Kim

Compressive Failure Threshold of Brittle Materials. 131
E.B. Zaretsky, V.E. Paris, G.I. Kanel, A. Rajendran

Dynamic Response of B_4C-SiC Ceramic Composites 147
S. Hayun, N. Frage, M.P. Dariel, E. Zaretsky, and Y. Ashuah

Index. 157

Preface

Reliable ballistic protection of military and police personnel, equipment, vehicles, aircraft and helicopters is presently impractical without the use of ceramic-based armor systems. The development and manufacturing of ceramic armor and armor systems, as well as the study of fracturing of ceramics under ballistic impact and ballistic performance of armor systems, have significant attention by both ceramic manufacturers and military specialists during the last decade.

The International Symposium on Ceramic Armor and Armor Systems was held during the 107th Annual Meeting of The American Ceramic Society, April 10-13, 2005, in Baltimore, Maryland. This symposium brought together scientists and engineers working with the development, manufacturing and evaluation of armor ceramics and ceramic-based armor systems. A total of 33 papers, including 12 invited, were presented by leading specialists from 9 countries (Canada, Germany, Israel, Italy, Japan, Korea, Russia, Ukraine and the United States). The speakers represented universities, government research centers and laboratories and industry. The symposium attracted not only many ceramic specialists but also many armor design and ballistic specialists from different countries.

These proceedings contain 13 invited and contributed papers presented and discussed at the symposium. The papers describe the results of the latest achievements in the area of ceramic armor systems for personnel, vehicular and structural protection. They are devoted to ceramic armor design and modeling, ceramic armor materials and composites development and manufacturing, physical properties and structures of armor ceramics, fracture mechanisms of armor ceramics and composites and ballistic testing and performance of ceramic armor systems. The papers also consider new tasks and approaches in the area of armor ceramics and systems. Each manuscript presented was reviewed in accordance with The American Ceramic Society's review process.

As the organizer of the symposium and the editor of these proceedings, I am grateful to the session chairs (Drs. Richard Bradt and Do Kyung Kim) and to all the participants for their contribution, cooperation, productive discussions, and time and effort. Thanks go to all the reviewers for their comments and suggestions and to the ACerS Meetings and Publications Department staff for their assistance. The financial support of ACerS is gratefully acknowledged.

It is my hope that this volume will be good addition to the past ACerS published literature related to ceramic armor. This volume should be of interest to the researchers and engineers working with all aspects of ceramic armor systems. The results described will help in the future development and implementation of advanced ceramic armor with improved performance and reliability.

Eugene Medvedovski

Ceramic Armor Development and Study

ADVANCED CERAMICS FOR PERSONNEL ARMOR: CURRENT STATUS AND FUTURE

Eugene Medvedovski
Umicore Indium Products
50 Sims Ave., Providence, RI 02909, USA

ABSTRACT

The use of advanced ceramics for personnel armor allows the defeating of the projectile and the ballistic impact energy dissipation providing adequate ballistic protection. The development of lightweight and inexpensive ceramic armor is under ongoing attention by both ceramic armor manufacturers and armor users. Some achievements in the development of ceramics for personnel armor are summarized. These ceramics include alumina ceramics with an Al_2O_3 content of 97-99.7 wt.-%, newly developed lower-weight alumina-mullite ceramics, new silicon carbide-based ceramics developed in the systems of SiC-Al_2O_3, SiC-Si_3N_4-Al_2O_3, as well as some reaction-bonded carbide-based ceramics studied jointly with some organizations. The main properties of the considered ceramics, which affect ballistic performance, are examined and analyzed as a function of composition and structure. Only a combination of all relevant physical properties and microstructure, including the ability to dissipate ballistic energy, as well as optimization of manufacturing processes, should be considered for proper selection and evaluation of armor ceramics. It has been demonstrated that not only dense homogeneous advanced ceramics, but also heterogeneous materials with optimal compositions and structures, have remarkable ballistic performance. Depending on requirements for ballistic protection, armor systems may be designed to various configurations and weights based on the most suitable ceramic materials selected for specific applications. Body armor plates based on the studied ceramics provide ballistic protection to NIJ Level III and IV, including for satisfactory multi-hit performance, depending on the type and thickness of ceramics and backing materials. The directions of the further improvements and developments of armor ceramics and systems with lower weight or/ and with increased performance for personnel protection are discussed.

INTRODUCTION

Advanced ceramics is one of the most important components of armor systems. Reliable ballistic protection for military and police from rifle threats is presently generally impractical without the use of ceramic-based armor systems. Advanced ceramics assist to defeat projectiles through the ballistic impact energy dissipation. The mechanisms of ballistic protection for ceramic and metal armor are significantly different. Metals absorb the energy of projectile by a plastic deformation mechanism. In the case of ceramics, the kinetic energy of the projectile is absorbed by a fracture energy mechanism. Usually ceramic armor systems consist of a monolithic ceramic or composite ceramic-metal body covered by ballistic nylon and bonded with a high tensile strength fiber lining such as Kevlar™, Twaron™, Spectra™ or fiberglass. Some soft metals (e.g. aluminum thin sheets) may be also used as a backing material. Upon impact of the bullet (velocity greater than 700-800 m/sec), the hard ceramic body used is cracked and broken, and the residual energy is absorbed by the soft reinforced backing material. This backing material also must support the post-impact fracturing of the ceramic body and the defeated bullet.

Among different structural ceramics, some types of oxide ceramics (mostly, alumina ceramics) and non-oxide ceramics (mostly carbides, nitrides, borides) are commonly used for ballistic protection. General properties of some armour ceramics are mentioned in the literature [1-10] and are summarized in Table 1. Despite elevated density (up to 3.95 g/cm^3), alumina ceramics are used for ballistic protection as they provide relatively high physical properties and performance, low cost and an ability to be manufactured using a variety of methods, e.g. slip casting, pressing, injection molding, without the use of expensive equipment, such as kilns with special protective atmospheres. By comparison, non-oxide dense ceramics such as boron carbide (B_4C), silicon carbide (SiC), silicon nitride (Si_3N_4), aluminum nitride (AlN) and some others, including the materials based on their binary systems, have high physical properties and relatively low density (excepting titanium diboride-based ceramics) that may be more beneficial for ballistic applications than alumina ceramics. However, these ceramics are usually manufactured by hot pressing that is expensive and not very productive. Although pressureless sintered materials, such as commercially produced SiC ceramics, are less expensive than hot-pressed, they are still relatively expensive because their manufacturing requires kilns with special controlled atmospheres and very high temperatures for sintering. Reaction-bonded silicon carbide (RBSC) and boron carbide (RBBC) and some other reaction-bonded carbide-based ceramics are considered as prospective materials for armor applications due to their relatively lower cost than hot pressed or pressureless sintered ceramics, high physical properties and an ability to manufacture relatively large sized products [11, 12]. These reaction-bonded materials also demonstrate better integrity for multi-hit situations than dense homogeneous carbide-based ceramics.

Ceramic-matrix composites also demonstrate a high integrity after ballistic impact due to their mechanical properties and impact energy dissipation ability. The following ceramic-matrix composites are mentioned as armor materials [3]: ceramic reinforced with whiskers or fibers, such as compositions of Al_2O_3/SiC_w, Al_2O_3/SiC_f or Al_2O_3/C_f, and ceramics/particulate-based compositions (TiB_2/B_4C_p, TiB_2/SiC_p). Cermets such as Lanxide™ composites based on silicon carbide infiltrated with aluminum, Ni/TiC, Al/B_4C_p and some others also demonstrate superior performance. The majority of these materials are hot-pressed, and, therefore, are relatively expensive. Although some metal-infiltrated composites, such as Lanxide™ SiC/Al composite, are not hot-pressed, their processing steps and equipment are relatively expensive, and they may be prone to problems in manufacturing. It should be noted that the majority of starting materials for ceramic-matrix composites, as well as for non-oxide armor ceramics, are relatively expensive, that additionally increases the cost of such armor.

The development of lightweight and inexpensive ceramic armor materials is under ongoing consideration by both ceramic armor manufacturers and armor users. In this paper, some achievements in the development of new or improved ceramics for armor applications, especially for personnel armor, are summarized. These ceramics are manufactured by slip casting technology based on specially optimized processing steps, including ceramic slurries preparation, casting into plaster moulds providing required shapes and sizes, drying and followed firing at relatively low temperatures (e.g. below

1550°C for alumina and some SiC-based ceramics) at proper firing conditions. Ceramic body armor plates with various configurations and dimensions up to 280x320 mm with thicknesses from 4.5 to 12.5 mm, including "completed" plates bonded with backing materials, are commercially produced.

ALUMINA ARMOR CERAMICS

Alumina armor ceramics manufactured on the commercial basis and studied in this paper include materials of AL97ML, AL98, AL98.5 and AL99.7 (the number denotes an approximate Al_2O_3 content) [8]. All the developed ceramics were formulated based on specially selected starting materials and sintering aids, which promote sintering at relatively low temperatures. All these ceramics are fully dense (water absorption is not greater than 0.02%). Phase composition and microstructure of the AL97ML, AL98 and AL98.5 ceramics are similar, and they consist of corundum grains (the major phase) bonded by a small amount of anorthite crystals and a silicate-based glassy phase. A small amount of mullite crystals is also present in the AL97ML ceramics. The AL99.7 ceramics consists of corundum grains bonded by tiny spinel crystals and a very small amount of a glassy phase formed in the presence of oxides-impurities.

Generally, the structure of the noted alumina ceramics is uniform and microcrystalline. The grain size of the alumina ceramics depends on the initial batch composition, initial particle size and particle size distribution of the starting alumina powders. Alumina powders with a smaller particle and median crystal size provide a fine-crystalline structure with a smaller grain size. The average corundum grains range from 1-3 μm for the AL99.7 (mostly isometric) to 3-6 μm (isometric) and (2-3)x(5-8) μm (short prismatic) for the AL97ML ceramics. A glassy phase is distributed uniformly between grains and, as expected, the amount of the glassy phase increases as the alumina content decreases.

Alumina-zirconia (AZ) ceramics based on a specially optimized ratio between alumina and partially stabilized zirconia does not have a glassy phase; zirconia grains with a size less than 1 μm are uniformly distributed between corundum grains with sizes of 1-2 μm. The zirconia phase probably inhibits the corundum grain growth during sintering.

All these microstructure features affect physical properties and ballistic performance of the ceramics. Physical properties depend on the Al_2O_3 content, the size and shape of corundum grains, the amount, composition and distribution of a glassy phase cemented the crystalline phase, the presence and composition of the "secondary" crystalline phases and closed porosity. They also depend on the "stressed conditions" at the boundary of the corundum grains and a glassy phase. These factors are governed by the wetting of alumina particles by a liquid phase and by the interaction between them during sintering, firing and cooling, as well as by the difference in thermal expansion between crystalline and glassy phases. The key properties of the studied ceramics are presented in Table 2. A proper selection of alumina starting materials that allows optimizing ceramic microstructure with desirable grain sizes and, therefore, physical properties may be considered as one of the important factors of the improvement of ceramics. Alumina ceramics with high corundum contents and a fine microstructure generally demonstrate higher values of mechanical properties, such as hardness, strength, sonic velocity and Young's modulus. Alumina-zirconia ceramics also demonstrates high hardness and strength as well as increased fracture toughness that is explained by the optimized ratio between corundum and tetragonal zirconia grains and their microcrystalline structure. The presence of mullite crystals reinforcing a glassy phase cementing the major corundum grains has a positive effect on mechanical properties of the alumina ceramics. Among the studied alumina ceramics, AL98 and AL97ML materials with relatively elevated average grain size, as well as AZ ceramics, have lower brittleness values. In general, brittleness of alumina ceramics is significantly lower than dense homogeneous carbide-based ceramics that may be considered as a positive factor of armor materials.

ALUMINA-MULLITE ARMOR CERAMICS

This type of ceramics is based on the proper ratio of two major phases such as alumina and lower-density mullite $3Al_2O_3.2SiO_2$ (specific gravity of alumina and mullite are 3.95 and 3.2 g/cm^3, respectively). Recently developed alumina-mullite ceramics (AM materials) also have a very uniform and fine-grained microstructure formed by corundum grains (a major phase) with uniformly distributed mullite grains cemented by an aluminosilicate glassy phase. The grain size of these ceramics is 2-4 μm (mostly isometric) with a presence of larger grains (up to 6 μm); smaller grains with a size of 0.5-1 μm are also noted. A reduction of density of alumina-mullite ceramics is exhibited to 3.5-3.56 g/cm^3. Alumina-mullite ceramics with additional zircon and zirconia phases (ZAS materials) also have microcrystalline structures but density of this type of ceramics is greater (3.6-3.75 g/cm^3, depending on compositions). Although the physical properties of these ceramics (Table 3) are not as high as alumina ceramics with the only major corundum crystalline phase and a low content of a glassy phase, alumina-mullite ceramics have brittleness values comparable with brittleness of high-alumina ceramics. Alumina-mullite ceramics AM2 with a lower content of an aluminosilicate glassy phase and alumina-mullite ceramics contained a relatively lower content of zircon and zirconia phases ZAS2 have a higher level of the physical properties. AM2 material may be considered as the most promising among others due to relatively high properties relevant to ballistic performance, lower brittleness and the lowest density (i.e. providing a lower weight of armor).

SILICON CARBIDE-BASED ARMOR CERAMICS

The developed armor silicon carbide-based ceramics are based on the formulations in the systems of $SiC\text{-}Al_2O_3$, $SiC\text{-}Si_3N_4\text{-}Al_2O_3$, SiC-Si, $SiC\text{-}Al_2O_3\text{-}Si$ and $SiC\text{-}Si_3N_4\text{-}Al_2O_3\text{-}Si$ with, or without, small amounts of sintering aids [12]. The batch compositions of the ceramics and the grades of raw materials used have been developed and selected to achieve an optimal particle size distribution and compaction, an optimal manufacturing ability and the occurrence of specific physical-chemical processes during firing conducted in special conditions.

The ceramics developed in the system of $SiC\text{-}(Si_3N_4)\text{-}Al_2O_3$ (AS and ASN materials) have a heterogeneous structure. It is formed by SiC grains with different sizes ranging from 3-5 to 120 μm bonded by a crystalline-glassy silicon carbide-aluminosilicate matrix. In the case of the use of Si_3N_4 constituent in the compositions (ASN materials), the bonding phase also includes silicon nitride and sialon. The compaction between SiC grains is relatively high, and it is achieved by specially selected ratios between the sizes of starting SiC particles. A high level of bonding between grains and the matrix is achieved due through a reaction-bonding mechanism. This reaction bonding occurs due to partial oxidation of some ingredients and the following high-temperature interaction of alumina with the products of the noted oxidation process. The structure in general and the bonding phase in particular are denser in the case of the nitride-based bonding phase. The use of small amounts of specially selected inorganic additives may also promote the liquid phase formation and the interaction between phases during firing. These ceramics have some porosity, but the pore size is small (approximately several microns), and the pores are uniformly distributed. The majority of initial pores between grains, which occur at the green body formation stage, may disappear during the firing process when the liquid phase is formed and the mullite crystals are initiated and grown. The surface of the ceramics has a higher content of a glassy phase than subsurface areas. This is a result of the firing conditions and the features of the physical-chemical processes, defined by diffusion mechanism, which occur during firing.

The ceramics made in the starting systems of SiC-Si, SiC-Al_2O_3-Si and SiC-Si_3N_4-Al_2O_3-Si (ESS and ESAS materials) also have heterogeneous structures with SiC grains ranging from 3-5 to 120 μm. The bonding phase is represented by silicon nitride and, in the case of the Al_2O_3-constituent, by sialon with a presence of a small amount of residual silicon. The bonding phases in these ceramics are also formed due to reaction-bonding mechanisms occurred during nitridation. The ceramics also have small porosity with very fine pore sizes. The sialon-bonded silicon carbide ceramics (i.e. the ceramics based on the initial system of SiC-Si_3N_4-Al_2O_3-Si) have lower porosity than other materials from this group.

Density of the SiC-based ceramics from the system SiC-(Si_3N_4)-Al_2O_3 is relatively low, ranging from 2.7 to 3.2 g/cm^3, depending on the composition. The achievement of these low values is connected with some porosity of ceramics and, more importantly, with high-temperature chemical processes, which cause the formation of new crystalline phase (mullite) with significantly lower density than starting alumina (3.2 vs. 3.95 g/cm^3). It is noted that the compositions containing a Si_3N_4 constituent provide lower density and lower porosity of the ceramics. These materials have zero shrinkage, or even slight expansion, as the ratio between green and fired dimensions is in the range of 1.0-1.02, depending on compositions. The studied SiC-Si_3N_4 and SiC-SiAlON ceramics prepared via nitridation also have low density of 2.85-3.05 g/cm^3 and zero shrinkage dealt with a formation of new phases through reaction-bonding processes.

Some physical properties for selected experimental SiC-based ceramics are presented in Table 4. These selected materials are distinguished by specific SiC and Al_2O_3 contents, by the presence and content of Si_3N_4, by the SiC particle size distribution (i.e. by the ratio between SiC constituents with different particle sizes), by the use of sintering aids and by the firing conditions. Ceramics based on the starting systems of SiC-Si_3N_4-Al_2O_3 and SiC-Si_3N_4-Al_2O_3-Si have higher mechanical properties, including hardness, due to a higher level of the grain compaction and bonding, and they are more prospective for ballistic applications. These ceramics also possess relatively high sonic velocity despite their heterogeneous structure and small porosity and relatively high Young's modulus.

REACTION BONDED SILICON CARBIDE CERAMICS

Some RBSC ceramics manufactured by several ceramic producers have been studied as armor materials. All of them are dense (with practically zero open porosity), and they have heterogeneous structures formed by SiC grains of different sizes up to 50-70 μm bonded by fine SiC grains and residual silicon (approximately 10-12%). Their structures depend on the size of starting SiC materials, particle size distribution, the source of carbon (mostly, pyrolysed organic binders) and applied technology.

Physical properties of RBSC ceramics depend on their structure and phase composition, i.e. on the content of residual silicon as the weakest ingredient, compaction and grain size distribution and completeness of the β-SiC formation during siliconization. Some physical properties of the studied RBSC ceramics are performed in Table 5. Hardness of these materials is relatively high, but its values have significant difference if it is measured for the primary SiC grains and for the matrix. The much lower hardness of the matrix is a sequence of the presence of residual silicon and, in some cases, an incomplete formation of the SiC phase. However, if the area of a matrix is relatively small, i.e. if the RBSC ceramics was formed with the initial SiC grains of specially selected particle size distribution, hardness of the matrix cannot be tested. Elastic modulus of RBSC ceramics is relatively high, and it may achieve the values comparable for dense homogeneous SiC ceramics. However, their mechanical strength is significantly lower than of homogeneous SiC materials due to heterogeneous structures, a presence of residual silicon and due to the stresses occurred during a new phase formation, as well as manufacturing process related stresses.

BALLISTIC PERFORMANCE

The fracturing process of ceramics during ballistic impact and penetration, when bullet velocities range from 700 to 5000 m/s, has several stages. As noted [3], this process includes: initial impact with hydrodynamic flow of penetrator and armor ceramics; breakup and continued flow of penetrator and high speed jetting of debris; ceramic fracture, formation of Hertzian cone cracks and tensile cracks on the back face with continued penetrator breakup and flow; erosion of penetrator and widespread fracture of the ceramics. At increased bullet velocities, the energy transmission through ceramic armor and across boundaries via shock waves becomes more valuable, i.e. the ability of ceramics to dissipate the bullet kinetic energy and to prevent the crack propagation is very important. During the ballistic impact, different kinds of cracks are formed resulting in the disintegration of the ceramic plate into particles ranging from a very fine powder to relatively large chunks (up to several square centimeters). The impact area of a ceramic is comminuted and compacted under the compressive action of the projectile.

Ballistic performance of ceramic materials depends on a number of properties. They include density and porosity, hardness, fracture toughness, Young's modulus, sonic velocity, mechanical strength, and some others. Any single property does not have a direct correlation with ballistic performance because the fracture mechanism during the bullet impact is very complicated, the crack formation is caused by different stress factors, and it occurs in an extremely short time. In short, the microstructural features affecting physical and ballistic properties strongly influence crack propagation and energy dissipation mechanisms and ultimately ballistic performance. Thus, the ceramic structures formed by the grains with different but specially selected sizes or formed by two or more crystalline phases positively affect ballistic performance. Hence, all relevant properties, as well as ceramic microstructural features, must be carefully considered in assessments of ballistic performance of protective systems. For instance, ceramics should have high hardness values in order to defeat a projectile and to decrease its velocity, but the crack propagation in ceramics after impact should not be intensive. In many cases, it is not easy to maintain these both conditions. Brittleness factor defined by hardness, fracture toughness and Young's modulus may be considered as one of the key factors affecting ballistic performance. It is, therefore, difficult to formulate properties requirements for all types of armor ceramics due to their significant difference of fracturing and energy dissipation during ballistic impact. It is more realistic to propose these requirements for each major group of armor ceramics. Some properties requirements for alumina ceramics, dense homogeneous carbide-based ceramics, heterogeneous RBSC and some other ceramics were discussed in the related studies [3, 8, 9, 11, 12]. A semi-phenomenological criterion of evaluation of ballistic energy dissipation ability may be expressed by the formula: $D = 0.36(HV.c.E/K_{Ic}^2)$ [7], which demonstrates that a combination of physical properties of armor ceramics should be considered in their evaluation and selection. Taking into consideration the formula of brittleness $B = HV.E/K_{Ic}^2$, the criterion of ballistic energy dissipation ability may be expressed as $D = 0.36(B.c)$. However, since ballistic energy dissipation also depends on phase composition and structure of ceramics, the D-factor may be expressed as $D = B.c.S$ (HV – Vickers hardness, K_{Ic} – fracture toughness, E – Young's modulus, c – sonic velocity, S – "structural" factor) [13]. Comparing studied alumina and alumina-mullite ceramics, they have the same level of brittleness and ballistic energy dissipation ability despite the relatively higher individual physical properties of alumina ceramics. These formulae may be successfully applicable for dense homogeneous armor ceramics. For heterogeneous armor ceramics, such as SiC-based materials AS, ASN, ESAS [12], ballistic energy dissipation ability and ballistic performance depend, to a significant degree, on phase composition and structure of the ceramics and related crack propagation.

All the studied ceramics have demonstrated a high level of ballistic performance. Armor systems based on the alumina, alumina-zirconia and alumina-mullite ceramics bonded with appropriate aramid-based and polyethylene backing materials are capable of defeating 5.56x45-mm SS109, 7.62x39-mm Russian and 7.62x51-mm NATO Ball FMJ, 7.62x54R LPS and 7.62x63-mm AP M2 ammunitions. They provide ballistic protection to NIJ Level III or Level IV depending on the type of ceramics and a backing material (i.e. Level IV in conjunction with a ballistic vest). Armor systems for personnel protection have satisfactory multi-hit performance (up to 6-7 impacts to one body-armor plate). Depending on the ammunition, each material has its own features of fracturing connected with its physical properties and microstructure. Trauma for personnel armor plates made from these materials occurred at acceptable levels (i.e. not greater than 44 mm deformation in accordance with NIJ Standards). The alumina ceramics with higher hardness demonstrated less trauma and bullet intrusion. However, in this case, a greater degree of a crack growth is observed, probably due to lower brittleness values. As mentioned above, the penetrator is distorted and eroded during the initial contact with the ceramic and the erosion of the penetrator is greater as the hardness of the ceramic increases. As predicted, the highest level of bullet erosion was observed for the hardest ceramics, such as alumina-zirconia and AL99.7 alumina ceramics, which correlates well with the data [6]. These materials perform better against armor piercing projectiles with the highest hardness and kinetic energy, such as with a WC core. Different kinds of cracks may be formed during the ballistic impact. A locus of conoidal coaxial cracks starts at the impact point; while radial tensile cracks are initiated at the back surface close to the axis of impact. Star cracks are formed at the side of conoids. Tangential spall cracks occur due to shear stress waves reflected from the edges of a plate and due to the formation of the cone cracks; lateral spall cracks may also form due to the longitudinal stress waves reflected from the backing support. Fragments of damaged ceramics with different sizes ranging from big chunks to a fine powder are noted after fracturing. This observation is inherent to dense alumina, alumina-zirconia and alumina-mullite ceramics with lower brittleness. The relatively high ballistic performance of AM2 ceramics, despite their lower level of mechanical properties, may be explained by high energy dissipation ability and lower brittleness provided by a presence of two major crystalline phases (corundum and mullite).

Studied heterogeneous SiC-based ceramics also demonstrate high ballistic performance. Armor systems based on these ceramics are capable of defeating 5.56x45-mm SS109, 7.62x39-mm and 7.62x51-mm Ball FMJ and 7.62x54R LPS ammunitions providing ballistic protection to NIJ Level III with high multi-hit performance (up to 6 hits to one plate). Ballistic test results also confirmed the importance of microstructure and ballistic impact energy dissipation ability of ceramics on ballistic performance. During ballistic impact, the silicon carbide grains "stop" a high-velocity projectile (i.e. decrease its velocity significantly). The crystalline-glassy phase is fractured, but the propagated cracks stop at the hard silicon carbide grains surface or on the pore surface. The silicon nitride- and sialon-containing bonding phase reduces the crack propagation. In the case of RBSC ceramics, the fracturing occurs through residual silicon as the weakest and the most brittle phase. The damage zone is characterized by large and small chunks and powder. However, the balance of the ceramic plate is relatively intact, and it can provide further ballistic protection relatively close to the initial impact. The damage zone has a traditional conical shape with a locus of conoidal cracks initiated at the impact point and radial cracks initiated at the back surface. Other kinds of cracks (such as spall cracks) are also present. The bonding phase composition may have a key role in fracturing and ballistic performance. For example, if the bonding matrix in RBSC ceramics has lower hardness, i.e. if the level of "bonding" is not high enough (it may happen either in the case of a lower content of the formed β-SiC or in the case of an excessive content of residual silicon), the RBSC ceramics have elevated fracturing and elongated cracks upon ballistic impacts and, as a result, lower ballistic performance, especially in the case of multi-hit situations.

In general, projectile energy is decreased by approximately three times within 20 μsec of projectile impact [1], and the rate of the loss of energy and, therefore, ballistic performance, depend on the loss of a projectile mass and the loss of a projectile velocity. The loss of projectile mass occurs when the projectile is defeated by high-hardness ceramics and is eroded due to a high-friction effect. The coarse SiC grains with a size of 50-120 μm used in the studied ceramic compositions have a great abrasiveness, and they erode the moving projectile significantly. For example, projectiles such as 7.62x51-mm NATO Ball FMJ and 7.62x54R LPS after ballistic impacts on the studied SiC-based ceramics had high damage and erosion that is especially observed in the case of the SiC-Al_2O_3-Si_3N_4-based ceramics with a high percentage of coarse SiC-grains in the composition. It is difficult to evaluate the decrease of projectile velocity when it moves through a ceramic armor system with negative acceleration. However, it may be postulated that high-abrasive large-sized particles, such as the silicon carbide grains, promote remarkable friction between the moving projectile and the mentioned grains and, hence, may promote the decrease of a projectile velocity more efficiently than the fine-sized particles used for pressureless sintered ceramics. The characteristics of fracturing and crack propagation for these types of SiC-based ceramics (AS, ASN, ESAS) are similar to RBSC ceramics, and they distinguish from hot-pressed or pressureless sintered dense microcrystalline B_4C and SiC armor ceramics. Despite high hardness and other mechanical properties, mentioned homogeneous carbide ceramics demonstrate intensive shattering, complete disintegration of the impact zone and intensive crack propagation. Although they can defeat NATO Ball and AP ammunitions, these homogeneous carbide materials are not well suited to multi-hit applications. However, the studied heterogeneous SiC-based ceramics, even RBSC ceramics, are not as superior against AP projectiles with a WC core as the studied alumina, alumina-zirconia and alumina-mullite ceramics as well as hot-pressed B_4C ceramics.

CERAMIC ARMOR SYSTEMS

Ceramic armor systems may be designed taking into consideration several different factors, including the properties of the armor components. The nature and thickness of backing materials may have a significant influence on crack propagation due to their own abilities to reduce the stress. Ceramic plates with thicknesses of 7-9 mm bonded with aramid fabrics, such as Kevlar or Twaron (the number of aramid fabric layers depends on the ballistic requirements, type of fabrics, and ceramic performance) can stop a variety of projectiles. In general, the thinner the ceramic plate, the greater the number of aramid fabric layers or other backing materials is required. Practical experience shows that a decrease of a ceramic plate thickness of 0.5-1 mm, when the thickness ranges from 5 to 7 mm, may require a significant increase of aramid backing. The systems based on 5-5.5-mm ceramic plates (all types of ceramics) require thick aramid backing. In this case, aramid backing made of flexible fibers with a small diameter of filaments (the materials, such as Kevlar 129 or KM or Twaron T-flex with 1000-600 denier vs. to Kevlar 29) are preferable. Completed systems made of thin ceramic plates (up to 5.5 mm) and aramid backing usually cannot defeat 7.62x63-mm AP rounds. Opposite to them, the systems based on thin ceramic plates with a thickness of 4.5-5.5 mm bonded with laminated polyethylene backing (20-25 mm thickness) have been shown to successfully defeat armor piercing projectiles with a WC core. This is due to the general high ballistic performance of laminated polyethylene (it can defeat 7.62x51-mm NATO Ball FMJ, but has not been shown to defeat 7.62x63-mm AP M2) as it locally softens and melts upon ballistic impact. In all cases, wrapping or confinement of ceramic plates with special grades of fiberglass or some other materials generally significantly increase ballistic performance. In many cases, more complicated designs based on the face ceramics and backing made from different materials (e.g. aluminum plus Kevlar or some other combinations) increases performance; however, it may decrease manufacturing productivity and, hence, increase the cost of the armor. The use of "stronger" ceramics (e.g. with higher hardness and strength but with a relatively low level of brittleness), when ceramic plates have 7-9-mm thicknesses, may allow for a decrease in the backing or a slight decrease of ceramic plate thickness. For example, in some cases, the use of alumina ceramics AL99.7, instead of AL98, provided such a benefit.

The armor design providing a decrease of the breakage of ceramics from the reflection of the shockwaves from the backing may be very effective. The armor plates consisting of a thin ceramic plate, aramid backing and a lightweight layer based on the special polymer-ceramic mix placed between the plate and the backing demonstrated high ballistic multi-hit performance (e.g. they can resist to 3 rounds of NATO Ball or LPS ammunitions). Such design made of AM2 plates with a thickness of 4.5 mm provided a total weight of only 2100-2200 g. A similar design based on the plates with a thickness of 7-9 mm required a significantly lower thickness of backing in order to achieve acceptable ballistic performance.

In order to increase spall protection at the ballistic impacts, a special elastic polymer layer placed on the front of the armor plates may be very effective. However, some ceramic materials, such as silicon carbide-based ceramics AS and ASN, demonstrated high spall protection performance even without a use of special front polymer layers. Their superior performance comparatively with dense homogeneous and even with RBSC and RBBS ceramics may be dealt with the heterogeneous structures of ASN and AS materials formed by coarse SiC grains cemented by a crystalline-glassy bonding phase.

DIRECTIONS OF THE FURTHER DEVELOPMENT OF ARMOR CERAMICS

Further improvements and developments of armor ceramics and systems for personnel protection are mostly focused on a decrease in weight and an increase in the performance of the armor. Such directions include the development of monolithic materials with improved fracture energy dissipation by composition and structure optimization, ceramic composites (e.g. ceramic-ceramic- and ceramic-metal-matrix composites), which may increase the "ductility" of materials and decrease crack propagation, composite armor systems (e.g. laminated or monolithic ceramic-particulate systems), as well as ceramic-based armor systems with a proper backing of optimized design. Structural optimization may be achieved by the use of specific particle size distributions of starting materials, by applying specific directional crystallization processes or by the development of multiphase ceramic materials. The latter may demonstrate increased ballistic energy dissipation. Some of these methods have been successfully used for the development of the armor ceramics described above.

Armor monolithic materials with lower weight may include carbide- and nitride-based ceramics with two, three or even more crystalline components (e.g. B_4C-SiC, SiC-Si_3N_4, SiC-AlN, SiC-SiAlON and some other types of ceramics). They may be either homogeneous fine-crystalline or heterogeneous reinforced by whiskers, platelets, nanotubes or even by grains of different sizes with specially selected particle size distributions. The latter group of materials may provide not only a weight decrease in the armour, but also higher impact energy dissipation that will result in better multi-hit performance. The development of alumina ceramics with a significant amount of a second crystalline phase with lower specific gravity also may be considered as a potential direction to achieve weight reduction of armor. The alumina-mullite ceramics described are a good example of a successful solution in this direction.

Search of inexpensive forms of raw materials for ceramics is another direction of the armor development. For example, some semi- or by-products from the alumina and aluminum industry may be explored as sources for alumina ceramics. These materials may be highly active in sintering and provide fine-crystalline structure and, therefore, may yield high mechanical properties. Non-traditional starting materials, contained the source of Si, C and some other elements, may be of interest for some armour ceramics. For instance, RBSC ceramics based on biomorphic processes may have a high potential for future material developments.

Further improvement or development of new manufacturing processes may also be considered as an important direction of armor development. An increase of productivity and output and a decrease of production cost are under ongoing consideration of many manufacturing companies. New processes may include pressure casting technology, presently mostly used for sanitaryware, rolling technology, processes focused on laminated structures manufacturing, new methods of infiltration of porous ceramic performs. Some of these new processes may have a high potential by providing high productivity and lower cost. Reaction-bonding processes may also have a significant role in the formation of armor ceramics with high multi-hit performance and relatively lower costs. The materials described in [11, 12] are good examples of this approach.

REFERENCES

[1]. C.F. Cline, M.L. Wilkins, "The Importance of Material Properties in Ceramic Armor"; pp 13-18 in *DCIC Report 69-1*; Part I: "*Ceramic Armor*", 1969.

[2]. Soon-Kil Chung, "Fracture Characterization of Armor Ceramics", *American Ceramic Society Bulletin*, **69** [3] 358-66 (1990).

[3]. D.J. Viechnicki, M.J. Slavin, M.I. Kliman, "Development and Current Status of Armor Ceramics", *American Ceramic Society Bulletin*, **70** [6] 1035-39 (1991).

[4]. I.Yu. Kelina, Yu.I. Dobrinskii, "Efficiency of the Use of Silicon Nitride Ceramics as an Armor Material" (in Russian), *Refractories and Technical Ceramics*, [6] 9-12 (1997).

[5]. B. Matchen, "Application of Ceramics in Armor Products"; pp 333-342 in Key Engineering Materials, Vol. 122-124, *Advanced Ceramic Materials*. Edited by H. Mostaghasi. Trans. Tech. Publications, Switzerland, 1996.

[6]. R.G. O'Donnell, "An Investigation of the Fragmentation Behaviour of Impacted Ceramics", *Journal of Materials Science Letters*, **10**, 685-88 (1991).

[7]. V.C. Neshpor, G.P. Zaitsev, E.J. Dovgal, et al., "Armour Ceramics Ballistic Efficiency Evaluation"; pp 2395-401 in *Ceramics: Charting the Future*, Proceedings of the 8th CIMTEC (Florence, Italy, 28 June-4 July 1994). Edited by P. Vincenzini, Techna S.r.l., 1995.

[8]. E. Medvedovski, "Alumina Ceramics for Ballistic Protection", *American Ceramic Society Bulletin*, **81** [3] 27-32 (2002), [4] 45-50 (2002).

[9]. B. James, "Practical Issues in Ceramic Armour Design"; pp 33-44 in *Ceramic Armor Materials by Design*, Ceramic Transactions, Vol. 134. Edited by J.W. McCauley, A. Crowson, W.A. Gooch, Jr., et al., 2002.

[10]. B.A. Galanov, O.N. Grigoriev, S.M. Ivanov, et al., "Structure and Properties of Shock-Resistant Ceramics Developed at the Institute for Problems in Materials Science"; pp 73-81 in *Ceramic Armor Materials by Design*, Ceramic Transactions, Vol. 134. Edited by J.W. McCauley, A. Crowson, W.A. Gooch, Jr., et al., 2002.

[11]. M.K. Aghajanian, B.N. Morgan, J.R. Singh, et al., "A New Family of Reaction Bonded Ceramics for Armor Applications"; pp 527-539 in *Ceramic Armor Materials by Design*, Ceramic Transactions, Vol. 134. Edited by J.W. McCauley, A. Crowson, W.A. Gooch, Jr., et al., 2002.

[12]. E. Medvedovski, "Silicon Carbide-Based Ceramics for Ballistic Protection"; pp. 19-35 in *Ceramic Armor and Armor Systems*, Ceramic Transactions, Vol. 151. Edited by E. Medvedovski, 2003.

[13]. E. Medvedovski, P. Sarkar, "Indentation Testing of Armor Ceramics"; pp. 589-596 in *Ceramic Engineering and Science Proceedings*, Vol. 25, N.3. Edited by E. Lara-Curzio and M.J. Readey, 2004.

Table 1. Some Armor Ceramics and Their Properties

Ceramics	Density, g/cm^3	Vickers Hardness*, GPa	Fracture Toughness K_{Ic}, $MPa.m^{0.5}$	Young's Modulus, GPa	Sonic Velocity, km/sec	Flexural Strength, MPa
Alumina, sintered	3.60-3.95	12-18	3.0-4.5	300-450	9.5-11.6	200-400
Alumina-Zirconia, sintered	4.05-4.40	15-20	3.8-4.5	300-340	9.8-10.2	350-550
Silicon Carbide, sintered	3.10-3.20	22-23	3.0-4.0	400-420	11.0-11.4	300-340
Silicon Carbide, hot-pressed	3.25-3.28	20	5.0-5.5	440-450	11.2-12.0	500-730
Silicon Nitride, hot-pressed	3.20-3.45	16-19	6.3-9.0	-	-	690-830
Boron Carbide, hot-pressed	2.45-2.52	29-35	2.0-4.7	440-460	13.0-13.7	200-500
Titanium Diboride, sintered	4.55	21-23	8.0	550	-	350
Titanium Diboride, hot-pressed	4.48-4.51	22-25	6.7-6.95	550	11.0-11.3	270-700
Aluminum Nitride, hot-pressed	3.20-3.26	12	2.5	280-330	-	300-400

*Tested at different loads

Table 2. Some Physical Properties of the Studied Alumina and Alumina-Zirconia Armor Ceramics

Property	AL97ML	AL98	AL98.5	AL99.7	AZ
Density,* g/cm^3	3.74- 3.76	3.78-3.82	3.81-3.84	3.90-3.91	4.35-4.39
Young's Modulus, GPa	280-300	325-360	370-420	400-450	310-340
Sonic Velocity, km/s	9.5-9.9	10.0-10.5	10.6-11.3	10.7-11.6	9.8-10.0
Vickers Hardness HV10	1230-1260	1220-1330	1320-1420	1520-1560	1520-1580
Fracture Toughness K_{Ic}, $MPa.m^{1/2}$	3.0-3.3	3.2-3.3	3.3-3.4	3.1-3.4	3.9-4.0
Flexural Strength, MPa	-	250-350	270-360	320-380	500-560
Brittleness, Bx10^{-6}, m^{-1}	340-380	370-430	420-460	525-545	280-330
Ballistic Energy Dissipation Criterion Dx10^{-12}, 1/s (calculated)	1.70-1.95	1.50-1.60	1.80-1.95	2.20-2.40	1.15-1.20

* Water absorption is not greater than 0.02%

Table 3. Properties of Selected Alumina-Mullite Ceramics

Property	AM2	AM5	ZAS2	ZAS3
Density, g/cm^3	3.52-3.56	3.52-3.55	3.70-3.75	3.58-3.62
Vickers Hardness HV10, kg/mm^2 GPa	1130 11.1	1030 10.1	1180 11.6	940 9.2
Rockwell Hardness HRA	90.5	89	90	85
Fracture Toughness K_{Ic}, MPa.m$^{1/2}$	2.54	2.32	3.03	3.08
Sonic Velocity, km/s	8.62	8.76	9.04	8.38
Young's Modulus, GPa	237	241	275	228
Flexural Strength, MPa	350	300	350	275
Brittleness, Bx10^{-6}, m^{-1}	408	452	347	221
Ballistic Energy Dissipation Criterion Dx10^{-12}, 1/s	1.3	1.4	1.1	0.7

The values of properties presented are average

Table 4. Some Physical Properties of the SiC-Based Heterogeneous Ceramics

Property	Ceramics in the group SiC-(Si_3N_4)-Al_2O_3	Ceramics in the group SiC-Si_3N_4-Al_2O_3-Si
Density, g/cm^3	2.7 – 3.2	2.85 – 3.05
Rockwell Hardness HRC HRA	40 – 55 58 - 77	45 – 55 68 – 78
Flexural Strength, MPa	105 - 155	120 – 140
Impact Strength, kJ/m^2	1.86 – 2.24	-
Young's Modulus, GPa	240 - 310	260 – 280
Sonic Velocity, km/s	9.8 – 11.2	9.8 – 10.05

Table 5. Some Physical Properties of RBSC Ceramics

Property	HI	M3	MC
Density, g/cm^3	3.03-3.05	3.0	3.04-3.08
Hardness HK1, GPa, major grains	20.3	21.2	19.8-19.9
Hardness HK0.5, GPa, matrix	14.2	12.3	-
Hardness HV1, GPa, major grains	23.5-23.9	23.6	22.8-24.4
Hardness HV0.5, GPa, matrix	14.4-14.5	-	-
Fracture toughness K_{Ic}, MPa.m$^{1/2}$, major grains	2.16-2.38	2.51	2.4-2.8
Sonic velocity, km/s	11.6	10.3	10.5-11.3
Young's modulus, GPa	390-400	305	330-375
Flexural strength, MPa	240	-	190

Hardness of the matrix for MC ceramics could not be tested due to the structural features of the ceramics

Characterization of ALON™ Optical Ceramic

Thomas M. Hartnett
Surmet Corporation
33 B Street, Burlington
MA USA 01803;

Charles T. Warner
Surmet Precision Optics
41618 Eastman Drive, Murrieta
CA USA 92562

Donald Fisher
OpticsSolutions Company
PO Box 340038, Dayton
OH USA 45434

Wayne Sunne
Raytheon Missile Systems
1151 E. Hermans RD
Tucson, AZ 85706

ABSTRACT

ALON™ Optical Ceramic is a durable window material for UV, Visible and Mid IR window and dome applications. The mechanical, thermal, and optical properties of ALON products produced commercially by Surmet Corporation have been measured and this new data will be presented. Comparisons to previously measured data will be made.

Optical quality, low scatter ALON having high strength that is nearly double previously reported has been made. Average strength values of 700 MPa at 21°C and 631 MPa at 500°C have been measured for ALON specimens prepared by precision surface finishing techniques. Polished optical domes tested have survived severe thermal shock tests. These strength levels are comparable to those for single crystal sapphire. Strength, thermal conductivity, thermal expansion, refractive index, emissivity and absorption coefficient will be presented.

1.0 INTRODUCTION

Aluminum oxynitride, ALON is a transparent polycrystalline ceramic which has high strength and hardness. It is transparent from ultraviolet (UV) wavelengths to mid infrared wavelengths (MWIR). High speed IR missile dome applications require a durable transparent material that can withstand severe thermal shock due to aerodynamic heating, remain transparent throughout the entire flight trajectory and be highly resistant to rain and sand damage. Light weight transparent armor applications require high strength and hardness in the outer layers of a ballistic laminate to defeat high velocity projectiles. Lighter weight ballistic laminate designs can be achieved by using thinner ALON panels having greater fracture strength.

The technology for the making ALON transparent ceramics was transferred from Raytheon Company to Surmet Corporation in 2002. The manufacture of dozens of large ALON panels for armor performance evaluations and the production of hundreds of domes, lenses and point of sale (POS) scanner windows have been achieved at Surmet over the past two years. In light of this recent manufacturing scale-up of the production of ALON it is important to reevaluate the critical material properties such as strength, thermal conductivity, thermal expansion, and transparency and demonstrate that the material is as good as or better than that which was produced in a laboratory/pilot scale process.

The data and results presented in this paper are part of Air Force sponsored Phase II SBIR program structured to addresses the need for an up to date and accurate material properties database for currently produced transparent ALON. Results of these measurements will be used in the future to model and predict the performance of ALON in missile dome and transparent armor applications. This materials properties study also successfully demonstrated that the same deterministic grinding and finishing operations could be used to produce high strength test samples as well as low-cost high strength thermal shock resistant missile domes.

2.0 MECHANICAL PROPERTIES

Mechanical properties testing of ALON was performed at the University of Dayton Research Institute (UDRI) in Dayton, Ohio. Biaxial Flexural Strength (at 21°C and 500°C), Young's Modulus, shear Modulus and Poisson's Ratio were measured using standard ASTM C1499-03[1] and ASTM C1259-01[2] test methods. Weibull modulus and characteristic strength were determined by fitting the measured data to a standard two parameter Weibull distribution given by Equation 1, ASTM C 1239-94a [3, 4]. Figure 1 shows the test fixture used for the biaxial flexure strength testing at room temperature and 500°C. Equation 2 is used to calculate the strengths from the sample and fixture dimensions (see Table I) and the breaking load. Grafoil™ discs having a thickness of 0.005" thick were placed between the sample disc and the load and support rings to reduce the contact stresses that are thought to cause some samples to fail under the loading ring or support ring.

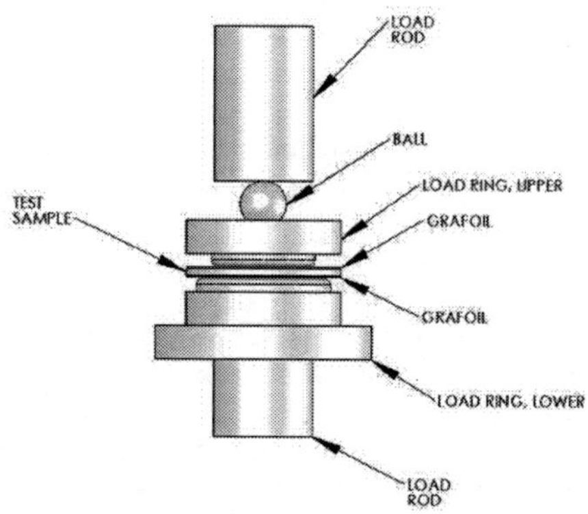

Figure 1: Biaxial Flexure Test Fixtures and Schematic of Test Setup used at UDRI for Testing ALON Samples. The loading rate used was 0.02"/min for all measurements; Humidity 50%; Temperature 21° C and 500 °C

Where: P_f = Probability of Failure
σ = Measured Strength
m = Weibull Modulus
σ_θ = Weibull Characteristic Strength

$$P_f = 1 - e^{-\left(\frac{\sigma}{\sigma_\theta}\right)^m} \quad (1)$$

Where: σ_f Breaking stress
F Breaking load
D_S Support Ring Diameter
D_L Load Ring Diameter
D Sample Diameter
h Sample Thickness
υ Poisson's Ratio

$$\sigma_f = \frac{3F}{2\pi h^2}\left[(1-\upsilon)\frac{D_S^2 - D_L^2}{2D^2} + (1+\upsilon)\ln\frac{D_S}{D_L}\right] \quad (2)$$

Table I: Sample Size and Biaxial Flexure Fixture Dimensions.

Sample Size and Test Fixture Dimensions				
Test Period	Sample Diameter	Sample Thickness	Load Ring Diameter	Support Diameter
(Historical) Raytheon 2002	0.998"	0.051"	0.4500"	0.9000"
(Surmet) May 2004	0.994"	0.055"	0.4166"	0.8330"
(Surmet) October 2004	1.244"	0.055"	0.4166"	0.8330"

Elastic Properties Measurements

The elastic properties of ALON were measured according to ASTM C1259-01 using the same disc specimens (1.25" diameter x 0.055" thick) used for biaxial flexure testing. The values measured at UDRI for the room temperature Young's Modulus, shear modulus and Poisson's ratio are given in Table II with comparisons to previously measured values. These current measured values are only slightly different from historical values [5, 6, 7, and 8]. Test specimens in all three set of measurements were 100% dense transparent ALON. The difference may be in the due to the greater accuracy of the flexural resonance method used in the current measurements.

Table II: Room Temperature Elastic Properties (E, G and υ) of ALON

Value	UDRI (2004)	Historical*(ref 5,6,7) SORI (1988)	Historical*(ref 8) Raytheon (1984)
Young's Modulus E (GPa)	321.05	321.3	323.8
Shear Modulus, G (GPa)	127.35	124.55	130.24
Poisson's Ration ,υ	0.26	0.24	0.24
* Historical (1988 and 1984) Data Measured in four point flexure using strain gages.			

Biaxial Flexure Strength Measurements

The testing program initially set out to compare three different grades of ALON and two test temperatures (RT and 500°C). The results of these measurements (May 2004) showed average strengths comparable to historical values (Historical Raytheon 2002) [9], however, the Weibull modulus was very low and the there were some very low and some very high strengths measured. Since the material is transparent it is relatively easy to inspect for the presence of strength limiting flaws. Having not found anything of significance the next step was to examine other possible reasons for the wide variation in strengths. There were some concerns that there was damage introduced into the samples by the fixed abrasive grinding process. Typically free abrasive grinding is used to lap the material prior to final finish. There were several other concerns which included the one-inch sample diameter, and the possibility of damage introduced during the edge beveling step.

It was decided that the next set of tests compare both fixed and free (loose) abrasive grinding as well as the benefits or detriments of chemical etching the ALON prior to mechanical testing. Chemical etching or a combination of chemical etching followed by a light polish has been demonstrated to reduce the average flaw size and increase the fracture strength of garnet (YAG, GSGG, and GGG) slab lasers [10]. Table III describes the complete set of samples and test conditions evaluated in this study.

Table III: Sample Set Descriptions

Sample Set	TEMP (°C)	RH	Number of Samples	Grain Size (μm)	Date Tested
Fixed No Etch RT	21	54%	14	233±29	9/13/04
Loose No Etch RT	21	48%	15	252±34	9/13/04
Combined No Etch	21	N/A	29		
Fixed/Loose No Etch	500	AIR	8	233±29	10/6/04
Fixed Etched RT	21	50%	16	233±29	9/14/04
Loose Etched RT	21	50%	14	252±34	9/14/04
HP ALON RT	21	49%	29	309±60	5/10/04
LS ALON RT	21	64%	30	254±43	5/27/04
HP ALON 500C	500	AIR	30	309±60	4/29/04
2002-RT	21	AIR	28	250	2002
2002-500C	500	AIR	31	250	2002

A tabulation of the measured strengths and Weibull parameters for the samples tested is shown in Table IV. A wide variation in strength is seen for the May 2004 data set. In this data set the room temperature average strength for HP-Grade ALON is 344 MPa. The high strength for this set is 746 MPa, a low strength is 185 MPa and Weibull modulus is 3.0. The failure probability distribution for this sample set shown in Figure 2 illustrates this wide distribution of strengths. A comparison of the May 2004 data with the Historical 2002 data shows that the average strengths are in the same range, 310 MPa to 390 MPa, but the Weibull Modulus is much lower and there are no extremely high or low strengths.

The high strengths measured in the first round of testing were very promising since there had never been strengths at this level ever reported for ALON transparent ceramics. The reason for low strengths was of concern and an investigation of the possible causes was begun. A low fracture strength sample is shown in Figure 3. The sample fractured into only about 6 pieces indicating a low amount of stored strain energy prior to fracture. The area where the fracture originated was located by examining the fracture pattern of the disc. The portion of the sample which contained the fracture origin was etched to reveal the grain structure and any surface defects. It can be seen in Figure 3 that there are deep scratches in the sample near the origin. These scratches were not visible using light microscopy prior to etching and indicate that there is damage in this sample which was most likely introduced during the grinding operation. This damage was not removed in subsequent processing steps and resulted in a failure at a much lower than average strength.

Table IV: Strength and Weibull Analysis for ALON Sample Sets

Sample Set	AVG	STDEV	Weibull Characteristic Strength	Weibull Modulus	Confidence Level
Surmet Sept.(2004)	σ (MPa)	(MPa)	σ_θ (MPa)	m	R^2
Fixed No Etch RT	700	169	750	8.5	0.982
Loose No Etch RT	753	179	811.9	5.6	0.968
Fixed No Etch 500°C	622	93	635.5	8.4	0.91
Fixed RT (light etch)	422	59	451	10.1	0.951
Loose RT (deep etch)	281	18	287.8	26.3	0.977
Surmet May (2004)					
HP ALON RT	344	146	361.6	3.0	0.953
LS ALON RT	389	135	425.2	3.2	0.988
HP ALON 500° C	364	123	410.4	3.0	0.990
Raytheon (2002)					
2002-RT	374.7	85.8	409.2	4.6	0.995
2002-500C	367.9	47.7	384.4	7.8	0.964

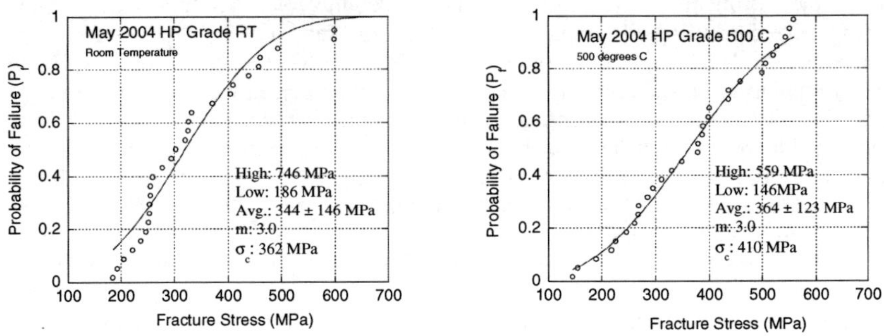

Figure 2: Weibull Failure Probability for HP-ALON Tested at RT and 500°C (May 2004)

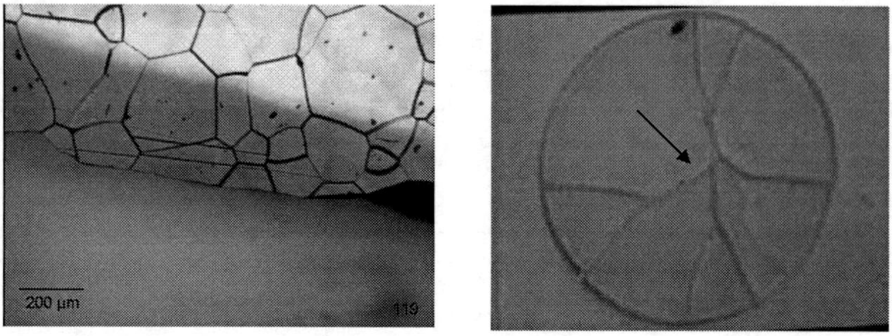

Figure 3: Photo of fracture test disc and an optical micrograph of the etched sample #119 near the fracture origin. Strength: σ_f = 197 MPa and Average Grain Size = 277 µm. Scratches on the sample near the fracture origin are revealed by etching.

Several changes were made to the samples fabricated for the second round of testing (September 2004). These changes were: (1) The sample diameter was increased from 1-inch to 1.25 inch to eliminated the possibility of premature edge related failures; (2) Samples were prepared by fixed abrasive and free abrasive grinding prior to final polish and the amount of material removed at each successive grit size was increased; and (3) A chemical-mechanical ("super" polish) final polishing step was used. Such techniques have recently been reported (reference 4) to result in high strengths for c-plane sapphire.

In addition, two sets of etched samples were prepared from the "super" polished sample groups. The etching was performed so that the effects of machining damage would be reduced. After polishing the samples were etched in 200°C phosphoric acid for two different time durations. The microstructure of polished and etched surfaces and the IR transmission before and after etching are shown in Figure 4. The light etching decrease the transmission in the IR only slightly and the deeper etch reduces the transmission by about 10%.

The strengths of the polished and etched ALON samples are lower than the strengths of the "super" polished samples. The average strengths decrease for the "super" polished and etched samples but the Weibull modulus increased. In the case of the lightly etched samples the average strength of 422 MPa is still higher than the average strengths reported for the May 2004 testing or for the Historical 2002 testing. The Weibull modulus for this lightly etched sample set is 10.1. The more deeply etched samples have an average strength of 281 MPa and a very high Weibull modulus of 26.3. So, unlike the data of Marion (1987) the strength of ALON is not increased by etching in phosphoric acid. The potential reliability, however, is increased. This increased reliability may not be practical without re-polishing because of the transmission decrease due to the increased surface roughness. One interesting observation is that there was no evidence of machining damage revealed by the etching process as was seen for the May 2004 test samples (see Figure 3). Many samples from that earlier May 2004 test group also show scratches revealed by light acid etching.

The Weibull failure probability distribution and representative samples from the Fixed Abrasive Group are shown in Figure 5. The average strength of this set of samples was 700 MPa and the Weibull modulus is 8.5. The highest strength sample in the fixed abrasive group had strength of 966 MPa and fractured into numerous small fragments. The zone under the load ring appears pulverized which is an example of a very high strength fracture. There is only one low strength sample in this group with the majority of the samples having strengths exceeding 600 MPa. Similar strength behavior was demonstrated for free abrasive ground ALON test at room temperature and both free and fixed abrasive ground ALON tested at 500°C. The average strength at 500°C is 622 MPa with a Weibull modulus of 8.4.

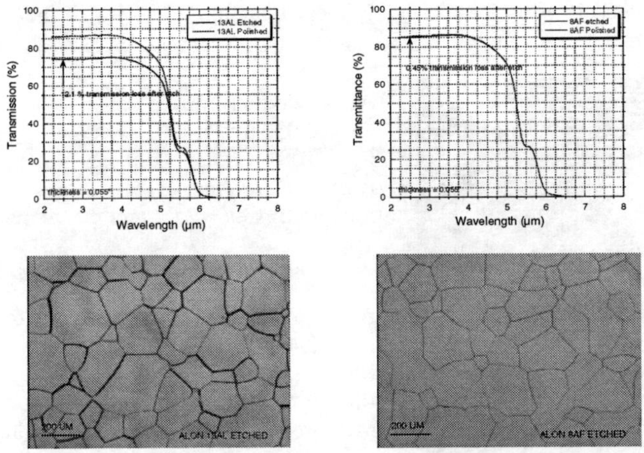

Figure 4: Microstructure and IR transmission of deeply etched (13AL) and lightly etched (8AF) ALON biaxial flexure test samples.

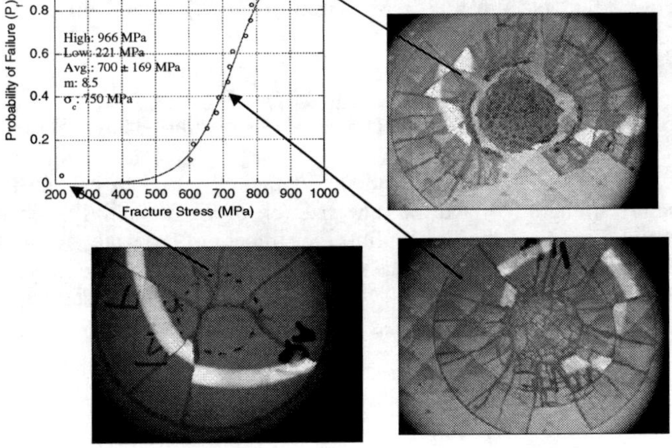

Figure 5: Weibull Failure Probability Distribution for LS-ALON (September 2004) Tested at Room Temperature

Surface roughness measurements using a ZYGO NewView 100 optical profilometer were measured for samples from the two etch groups and an un-etched sample from the high strength group. Figure 6 illustrates the nature of the surface roughness features and the level of surface roughness produced by the etching process. The most deeply etch surface sample has a roughness PV roughness of 5.0 μm and an Ra of 0.657 μm compared to a PV roughness of 2.44 μm and an Ra of 0.215 μm for the lightly etched sample. The Ra of the "super" polished samples is 10 Å and the PV roughness is 30 nm.

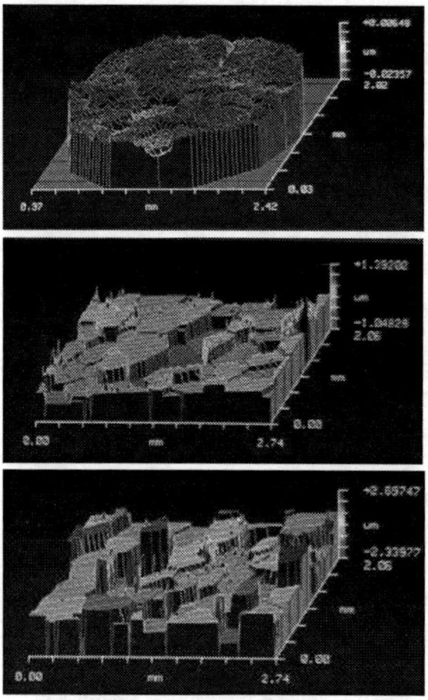

ALON Fixed Abrasive Polish
PV = 0.030 μm
Ra = 0.001 μm
RMS = 17.71 Å
Measured Area = 2.07 mm x 1.99 mm

ALON 8AF Fixed Abrasive Mild Etch
PV = 2.44 μm
Ra = .215 μm
RMS = .261 μm
Measured Area = 2.74 mm x 2.06 mm

ALON 13AL Loose Abrasive Deep Etch
PV = 5.036 μm
Ra = 0.657 μm
RMS = 0.887 μm
Measured Area = 2.74 mm x 2.06 mm

Figure 6: Surface roughness of polished and etched ALON biaxial flexure test samples.

The Griffith relationship given by Equation 3 commonly used to explain the relationship between the fracture strength (σ_f) and the flaw size (c_f) of a material. The fracture toughness (K_{1c}) of the material determines how susceptible the strength of the material is to the size of the flaw or crack. The factor (Y) is related to the shape and the location of the flaws of the material.[4, 11, 12, 13, 14, 33] For brittle ceramics and glass which fail in tension due to surface flaws the factor Y is equal to about 1. The fracture toughness (K_{1c}) of ALON is in the range of 2.0 to 2.4[8, 15].

$$\sigma_f = \frac{K_{1c}}{Y\sqrt{c_f}} \tag{3}$$

Table V compares the predicted flaw size calculated using equation 3 and the measured average strength with the peak to valley (PV) surface roughness and the grain size. There is no apparent relationship between the fracture strength and the grain size. There is a strong correlation between the surface roughness and the strength although there is not an obvious scaling factor to convert PV or Ra to the flaw size. The lower surface roughness "super" polished sample group has the highest strength and the increased roughness caused by etching lowers the strength accordingly. Based on the higher Weibull modulus the non-uniform etching rate of the adjacent ALON grains provides creates a more uniform distribution of flaws on the surface. The greater etch depth smoothes out the distribution of larger subsurface flaws introduced by machining and creates a new controlled distribution of larger flaws that lowers the average strength but increases the Weibull modulus. The high strength "super" polished and un-etched ALON has a low level of subsurface damage that is not revealed by etching as was the case for the low strength sample #119 and other in that group tested in May 2004. The nature of this subsurface damage that cannot be exposed by etching will need to be examined in greater detail using TEM and X-Ray topography which can reveal dislocations, twins and strains in the polished surfaces.[16, 17]

Table V: Predicted Flaw Size for ALON Fracture Test Samples Compared to the Surface Roughness

Sample Set	Measured Average Strength (MPa)	Weibull Modulus	Calculated Flaw Size (μm)	Measured PV Surface Roughness (μm)	Measured Grain Size (μm)
Polished Fixed Abrasive 9/2004	700 ±169	8.5	12	0.030	233±29
Mild Etch Fixed Abrasive 9/2004	422 ±59	10.1	32	2.44	233±29
Deep Etch Loose Abrasive 9/2004	281 ±18	26.3	73	5.04	233±29
Sample #119 from LS-Group	197(individual)		148	scratched	277±48
LS-Group May 2004 testing	389 ±135	3.2			254±43

3.0 THERMAL PROPERTIES

The thermal properties of transparent ALON have been evaluated in the past by several labs [7, 18] There is a need for more recent and up to date data measured for currently available transparent ALON. These material properties will be used in predictive models for thermal shock performance in high speed missile dome application. The measurements reported in this paper are more accurate than historical data and cover a wider temperature range. Samples were prepared using LS-Grade ALON (low scatter optical grade) produced at Surmet Corporation, Burlington, MA. The measurements of the thermal properties of ALON were made at TPRL, Inc. [19]. Thermal Diffusivity was measured in the temperature range -150°C to 700°C using the Laser Flash Diffusivity method (ASTM E1461). The sample was gold coated to minimize the effects of radiation heat transfer. The Specific Heat (c_p) was measured in the temperature range: -50°C to 700°C using Scanning Differential Calorimeter Perkin Elmer Model DSC-2, (STM E1269). Thermal conductivity was calculated from the bulk density (3.68 g/cc), the thermal diffusivity and the specific heat. The Thermal Expansion Coefficient (CTE) was measured in the temperature range -150°C to 800°C using a dual push-rod dilatometer (Theta Dilatronics II).

A summary of the thermal properties measured is listed in Table VI and comparison plots to historical data are shown in Figures 7 thru 9. Figure 7 shows that the thermal conductivity of ALON measured at TPRL is slightly higher in the temperature range from -50°C to 600°C than previously measured. The high temperature values (T > 600°C) reported previously are most likely in error due to the radiation effects of heat transfer through the transparent samples. In the current measurements the radiation heat transfer effects were minimized by applying a gold coating to both sides of the test sample.

Thermal Expansion values are slightly higher than previously reported and these new measurements extend the temperature range to include temperatures from -150°C to 800°C. A greater number of points were measured on the specific heat vs. temperature curve and the temperature range of the measurements was extended to include points below 0°C to -50°C.

Table VI: Thermal Properties of Transparent ALON Measured at TRPL (2004)

Temp.	MEAN CTE	Specific Heat	Thermal Diffusivity	Thermal Conductivity
°C	10^{-6} in/in - °C	J/g-K	cm^2/s	W/m-K
-150.0	3.067	-	-	-
-100.0	3.591	-	-	-
-50.0	4.070	0.600	0.0789	17.36
-20.0	4.336	0.687	0.0587	14.78
0.0	-	0.733	0.0500	13.44
23.0	-	0.781	0.0430	12.30
100.0	5.230	0.916	0.0312	10.48
200.0	5.783	1.015	0.0236	8.79
300.0	6.196	1.084	0.0201	7.98
400.0	6.507	1.132	0.0176	7.32
500.0	6.758	1.165	0.0161	6.89
600.0	6.986	1.191	0.0154	6.71
700.0	7.211	1.219	0.0137	6.12
800.0	7.432	1.235	0.0113	5.13

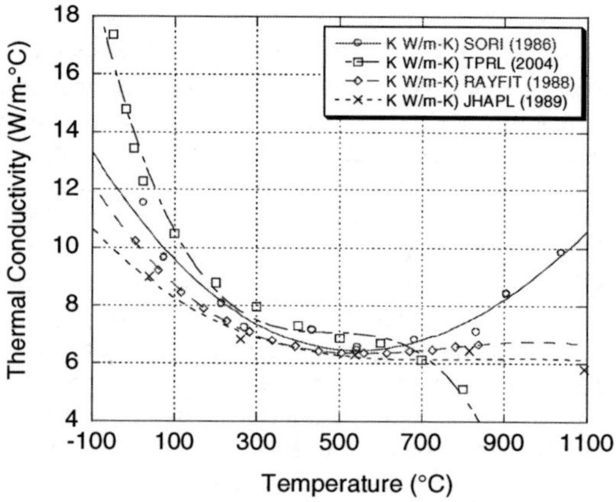

Figure 7: Thermal Conductivity of ALON – Comparison with Historical Data

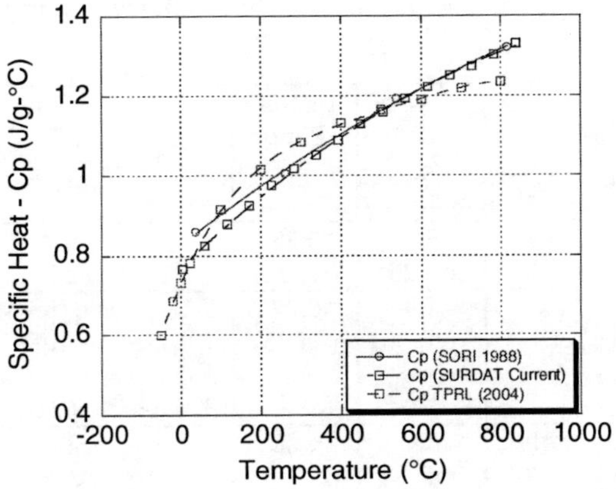

Figure 8: Specific Heat of ALON-Comparison with Historical Data

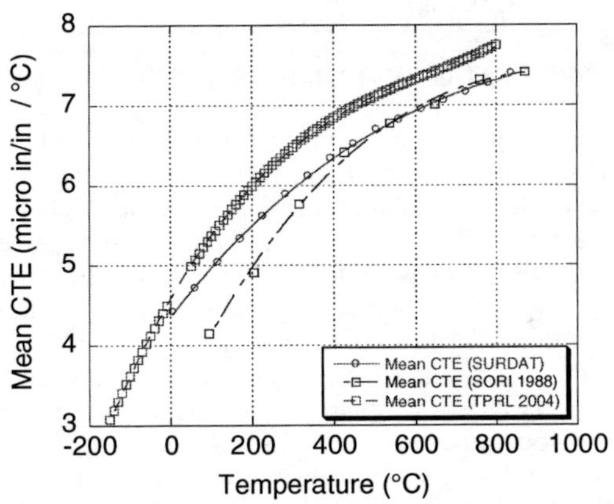

Figure 9: Thermal Expansion Coefficient of ALON--Comparison with Historical Data

The thermal shock figure (500°C) of merit, R' of ALON is compared to three other IR dome materials in Table VII. It can be seen in this table the predicted thermal shock resistance of ALON for "mild" thermal shock conditions is similar to mechanically polished sapphire, 2x that of spinel, and 4x that of MgF_2. Chemically polished sapphire (ref 4) has exceptionally high strength (1534 ± 378 MPa) and thus has a 3x lager predicted thermal shock resistance than ALON.

Table VII: Thermal Properties and Thermal Shock (@ 500°C) F.O.M Comparison Transparent Dome Materials

Material	C.T.E. α (10^{-6}/K)	Young's Modulus E (GPa)	Thermal Cond. κ (W/m-°C)	Poisson's Ratio ν	Strength σ_f (MPa)	Thermal Shock R' (W/m)
MgF_2	10.4 (ref 28)	114 (ref 32)	5.0 (ref 28)	0.28 (ref 32)	120 (ref 24)	365
Sapphire	7.70 (ref. 30)	345 (ref 33)	10.9 (ref. 30)	0.27 (ref 33)	1534 c-plane (ref 4) 708 c-plane (ref 4) 535 c-plane (ref 30) 375 c-axis 60° (ref 31)	4594 2120 1602 1123
ALON	6.76 (ref 19)	321 (ref 26)	6.9 (ref 19)	0.26 (ref 26)	622 (ref 26)	1464
Spinel	7.77 (ref 27)	193 (ref 32)	7.88 (ref 27)	0.26 (ref 27)	190 (ref 27)	739
Thermal Shock F.O.M.: $R' = \sigma_f (1-\nu)\kappa/\alpha E$						

4.0 THERMAL SHOCK PERFORMANCE (DOME TESTING)

Finished domes may have damage at the surface or contain internal flaws that are not visible or detectable by normal optical inspection methods. This damage can greatly reduce the strength of the domes and skew the predicted reliability that was based on Weibull statistical analysis of test coupons. Reliability testing of finished optical components such as domes is an important operation for screening out weak components. It is desirable that the method be a non-destructive method to screen domes for such defects and also provide a threshold which insures that only structurally robust domes make it to the missile level. The test must not weaken the component. Several methods have been proposed that would provide a non-destructive thermally or mechanically induced stress: quartz lamps, laser heating, hydraulic pressure and liquid quench have all been proposed or evaluated as suitable methods for proof testing domes.

Of these methods, the liquid droplet quench provides the thermal stress similar to that produced during actual flight and can be easily implemented as a 100% screen or proof test. Predictive models to determine the stress levels introduced by the liquid droplet quench test have been developed but are difficult to verify. The non-linear and highly dynamic phase change occurring as jets of water are impinged on the dome make analysis of the thermal contour and resulting stress field nearly impossible. It is only with the aid of experimental measurement using IR cameras and carefully time mapping individual elements that a relation between quench temperatures and peak dome stress can be ascertained.

While this is a rather laborious technique requiring extensive testing the test has been verified using sapphire domes having the same geometry as the ALON domes that were tested recently. The first order assumptions for the ALON dome test parameters were based on sapphire material property data. This assumes that the properties of ALON and sapphire of the materials vary similarly and do not significantly affect the stresses induced by the test.

Testing was performed on three ALON domes polished using the same grinding and polishing media and techniques as were used for the high strength biaxial flexure test specimens (σ_f =700

MPa ± 169 MPa, m = 8.5). The liquid droplet quench testing was performed at Raytheon's test facilities in Tucson, AZ. using the methods and parameters that were established for sapphire domes of the same geometry. ALON domes were heated to a uniform temperature of 150°C and water spray quenched on the concave surface to room temperature using a water temperature of 25°C. *All three domes survived this test.*

Testing of additional ALON domes is currently is process and predictive models of the thermal stresses generated during the quench testing are being refined using the newly measured temperature dependent thermal properties of ALON. The models will be used to alter if necessary the quench conditions so that the test conditions of liquid droplet reliability test for ALON domes can be finalized.

5.0 OPTICAL PROPERTIES

The purpose of this analysis is to quantify the emissive properties of ALON and compare these values with two other MWIR dome materials. This is important because a dome in flight will be aerodynamically heated and will begin to emit radiation to the detector behind the dome.

Several characteristic optical properties of ALON were measured and analyzed at JHAPL, Laurel, MD.[20] The data was further analyzed and reduced to final form by Optics Solution, Dayton, Ohio.[21] A brief summary of the data will be presented in this paper and a more complete description of the measurement methods and results can be obtained from Reference 22.

The ALON material measured was optical LS-grade ALON made by Surmet Corporation, Burlington, MA. LS-Grade ALON is large grained polycrystalline material having low optical scatter. The LS-Grade and HP-grade ALON are essentially the same in terms of purity and grain size. The primary difference is the optical scatter in the visible region of the spectrum. The spinel is polycrystalline optical grade material produced at Surmet Corporation. The sapphire was made by Crystal Systems and the comparisons made here are based on previous extensive measurements on sapphire made at JHAPL. Measurements of the transmission vs. temperature were made to obtain the extinction coefficient. The scatter (see Table VIII) at 3.39 μm was measured to determine the contribution of scatter to the extinction coefficient.

By knowing the index of refraction of the material and using the measured transmittance values, the total extinction coefficient can be calculated using equation 4.

$$\beta_{ext} = \frac{\ln\left(2R^2\tau\right) - \ln\left[\sqrt{(1-R)^4 + 4R^2\tau^2} - (1-R)^2\right]}{L}. \quad (4)$$

Where: $R = \left(\frac{n-1}{n+1}\right)^2$ is the single surface reflectance (5)

And: n is the index of refraction, L is the sample thickness, and τ is the transmittance.

The refractive index values for ALON (Reference 22) were computed using Equation 6.

$$n^2 = 1 + \frac{A\lambda^2}{\left(\lambda^2 - B\right)} + \frac{C\lambda^2}{\left(\lambda^2 - D\right)} \quad (6)$$

Where: A = 2.1578, B = 0.010507, C = 3.6683 and D = 243.02

The refractive index values for Spinel and Sapphire (Reference 24) were computed using the Equation 7.

$$n^2 = A + \frac{B\lambda^2}{(\lambda^2 - C)} + \frac{D\lambda^2}{(\lambda^2 - E)} \tag{7}$$

For Spinel A = 1.9370478, B = 0.95795641, C = 0.01832141, D = 3.7654327, and E = 300.

For Sapphire A = 1.6257584, B = 1.457221, C = .0113346735, D = 6.6760415, and E = 400.

The absorption coefficient is then calculated using the following relationship:

$$\beta_{ext} = \beta_{scatter} + \beta_{abs} \tag{8}$$

Table VIII: Total integrated scatter (TIS) and scatter coefficient (β).

Sample #	Sample Description	Wavelength (μm)	%Transmittance @ 3.39 μm	TIS	scatter coefficient □ (cm^{-1})
11	ALON (1" x 0.050")	3.39	86	0.005	
14	ALON (1" x 0.236")	3.39	88	0.011	0.0127
15	Sapphire (1" x 0.050")	3.39	88	0.005	
16	Sapphire (1" x 0.236")	3.39	87	0.012	0.0148
17	Spinel (1" x 0.080")	3.39	88	0.005	
18	Spinel (1" x 0.236")	3.39	87	0.02	0.0379

The temperature dependence of the absorption coefficient of ALON in the 3 to 5 μm wavelength range is shown in Figure 10. The absorption coefficient of spinel (Figure 11) and sapphire exhibit a similar temperature and wavelength dependence. A comparison of the calculated emittance behavior of these three materials is given in Table IX.

Table IX: Band averaged emittance of ALON, Spinel and Sapphire (3 μm to 5 μm)

	Band Averaged Emittance					
Material	25C	100 C	200 C	300 C	400 C	500 C
Spinel	0.024	0.025	0.029	0.037	0.045	0.064
ALON	0.076	0.087	0.104	0.122	0.144	0.179
Sapphire	0.044	0.048	0.058	0.071	0.087	0.107

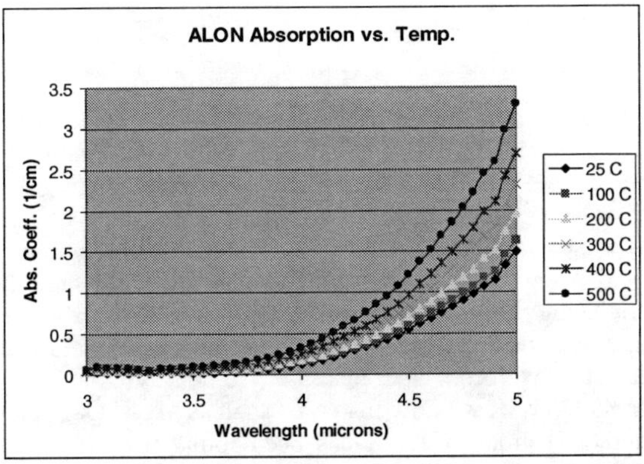

Figure 10: ALON: Absorption Coefficient vs. Wavelength and Temperature.

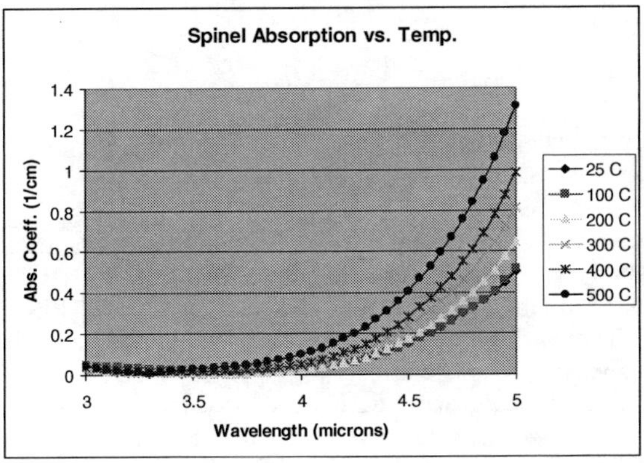

Figure 11: Spinel: Absorption Coefficient vs. Wavelength and Temperature.

A first order evaluation of the emittance is simply given by the equation below by Equation 9. The emittance is determined by the product of the absorption coefficient and the path length or sample thickness. Table 10 compares the band averaged emittance of ALON, spinel and sapphire domes having a thickness (L) of 0.090".

$$\varepsilon = \beta_{abs}L \tag{9}$$

The lower emittance of spinel makes it the most suitable optical material if dome emission is of primary concern. In general the emittance of ALON is about three times that of spinel and about slightly less than 2 times that of sapphire. The higher strength of ALON compared to spinel and its lower cost compared to sapphire and spinel makes ALON an attractive and viable dome material for high speed MWIR missile domes.

6.0 SUMMARY

Complete thermo-physical and optical properties of ALON have been tested for various grades of material produced using full-scale production processes. Complete data for the design of critical optical components incorporating ALON is given, including Weibull modulus and characteristic strength values. Deterministic grinding and polishing of ALON has resulted in characteristic strength values which are over 160% greater than that previously reported at high temperature (500C) and over 180% greater for room temperature samples. Through use of Surmet's proprietary finishing processes, the reliability of ALON has also increased significantly from previous reports. It is expected that the finishing techniques developed for ALON will also be useful for strength improvements and increased reliability of other hard polycrystalline ceramics and single crystal MWIR window materials such as $MgAl_2O_4$-spinel and sapphire.

ALON Low Scatter (LS) grade material has been found to have the best combination of mechanical and optical properties for use in severe environments. For thermal shock applications, ALON is comparable to Sapphire, while clearly superior to MgF2 and Spinel. Due to its lower cost for raw material and fabrication it will continue to replace sapphire in many applications.

REFERENCES

[1]ASTM C1499-03, "Standard Test Method for Monotonic Equibiaxial Flexural Strength of Advanced Ceramics at Ambient Temperature", (2003).

[2]ASTM C1259-01," Test Method for Dynamic Young's Modulus, Shear Modulus, and Poisson's Ratio for Advanced Ceramics by Impulse Excitation of Vibration", (2001).

[3]ASTM C1239-94a, "Standard Practice for Reporting Uniaxial Strength Data and Estimating Weibull Distribution Parameters for Advanced Ceramics", (1994).

[4]C.A. Klein, "Flexural Strength of Sapphire: Weibull statistical analysis of stressed area, surface coating, and polishing procedure effects", J. Appl. Phys. **96** [6] pp 3172-3179, (2004).

[5]M.W. Price, T.E. Hubbert, and J.R. Koenig, "Mechanical and Thermal Properties of Four IR Dome Materials", *Final Report to JHU/APL,* SRI , Birmingham, ALA. (1988).

[5]M.W. Price, T.E. Hubbert, and J.R. Koenig, "Mechanical and Thermal Properties of Four IR Dome Materials", *Report No. SORI-EAS-87-771-6225,* Interim Data Report to Johns Hopkins APL August (1987), Southern Research Institute, Birmingham, ALA. (1987).

[7]C.L. Resch, "Structural Limits of Sapphire, Zinc Sulfide, ALON, and Yttria IR Domes Based on Altitude and Mach Number", *Doc. No, AM-89-E0077*, JHAPL Laurel, MD (1989).

[8]RAYTRAN® ALON Property Data Sheet, Raytheon Company (1984).

[9]J.M. Wahl, "Biaxial Flexure Strength Testing of ALON™ Optical Ceramic", *Memo No. ME02-13*, Raytheon Company, MMEL, Advanced Materials, March (2002).

[10]J.E. Marion, "Fracture Mechanisms and Strengthening of Slab Lasers", SPIE Vol. 736 New Slab and Solid State Laser Technologies and Applications, pp2-12 (1987).

[11]B.R.Lawn, "Fracture and deformation in brittle solids: A perspective on the issue of scale", J. Mater. Res. **19** [1], pp 22-29 (2004).

[12]R.G. Munro and S.W. Freiman, "Correlation of Fracture Toughness and Strength", J. Am. Ceram. Soc. **82** [8], pp 2246-2248, (1999).

[13]J.E. Marion, D.M. Gualtieri, and R.C. Morris, "Compressive Epitactic layers on single crystal components for improved mechanical durability and strength", J.Appl.Phys.**62** [5] (1987).

[14]R.G. Munro, "Evaluated Material Properties for a Sintered α-Alumina", J. Am. Ceram. Soc. **80** [8], pp 1919-1928, (1997).

[15]J.J. Swab, G.A. Gilde, P.J. Patel, A. A. Wereszczek, J. W. McCauley, and J. D. Risner, " Fracture Analysis of Transparent Armor Ceramics", Fractography of Glasses and Ceramics IV, Ceramic Transactions, vol. 122, edited by J.R. Varner and G.D. Quinn, Westerville, OH, The American Ceramic Society, pp 489-508 (2001).

[16]D. Black, R.P. Polvani, L. Braun, B. Hockey, and G. White, "Detection of sub-surface damage in sapphire", SPIE Vol. 3060 Window and Dome Technologies V, pp. 102-114 (1997).

[17]F. Schmid, C.P. Khattak, S.G. Ivanova, D.M. Felt, and D.C. Harris," Influence of Polishing on the Biaxial Flexure Strength of Sapphire at 600°C", Proceedings of the 8th DOD Electromagnetic Windows Symposium April (2000).

[18]J.R. Koenig, "Thermal Conductivity of Spinel, ALON, and Yttria", *Report No. SORI-EAS-86-695-6081-I-F,* Final Report to Raytheon Company (July 1986), Southern Research Institute, Birmingham, ALA. (1986).

[19]J. Gembarovic, R.E. Taylor, and D.L. Taylor, "Thermophysical Properties of ALON", *Doc. No. TPRL 3199*, Report to Surmet Corporation September (2004), TPRL, Inc. West Lafayette, IN (2004).

[20]D.W. Blodget, D.V. Hahn, and M.E. Thomas, "Optical Characterization of SURMET samples", *Report For Order No. 542131, AD-26162, Surmet Corporation, Burlington MA* , Johns Hopkins APL, Laurel, MD (2004).

[21]D. Fisher, "Emissivity Evaluation of ALON™, Spinel and Al_2O_3 – Final Report", *Report Prepared for Surmet Corporation,* by Optics Solutions, Dayton, OH Jan 3, 2005.

[22]M.E. Thomas, D.W. Blodget, D.V. Hahn, "Optical Characterization of ALON™ and spinel", SPIE Vol. 5786 Window and Dome Technologies IX, (2005).

[23]D. Zelmon and D. Rush, "Refractive Index Measurements of ALON™ from 0.4 to 5.0 microns", *Unpublished Measurements Made for Surmet Corporation, Burlington, MA*, at AFRL, Materials and Manufacturing Directorate, WPAFB, Dayton, OH (2004).

[24]Handbook of Optics Vol II Devices Measurements and Properties, Michael Bass Editor in chief, McGraw Hill-Inc, , Chapter 33, pages 33.64 and 33.61 (1995).

[25]J A Savage, "Infrared Optical Materials and their Antireflection Coatings", Adam Hilger Pub., (1985).

[26]S. Goodrich, "Surmet ALON Equibiaxial Flexure Test", University of Dayton Res. Inst., Doc. No. SMG-04-2-59, (2004).

[27]J.R. Koenig, "Thermostructural Evaluation of Four Infrared Seeker Dome Materials", Part 2. Thermal and Mechanical Properties; *Report # NWC TP 6539*, Southern Research Institute (SORI), Birmingham, AL, (1985).

[28]Saint Gobain Product Data Sheet. Available at www.ndfc.saint-gobain.com

[29]Mellor Optics Product Data Sheet. Available at www.melleroptics.com

[30] "High temperature compression and ring-on-ring testing of sapphire" SPIE Vol. 3705 pg. 17-27, (1999).

[31]R.L. Gentilman, E.A. Maguire, H.S. Starrett, T.M. Hartnett, and H.P. Kirchner, "Strength and transmittance of sapphire and strengthened sapphire", J. Am. Ceram. Soc. 64 [9] C116-117, (1981).

[32]D.C. Harris, "Infrared Window and Dome Materials" SPIE Volume TT10 (1996)

SYNTHESIS OF DENSE B_4C-SiC-TIB_2 COMPOSITES

S. Hayun, N.Frage, H. Dilman, V. Tourbabin and M. P. Dariel
Department of Materials Engineering, Ben-Gurion University of the Negev,
P.0. Box 653
Beer-Sheva, Israel, 84105

ABSTRACT

The present communication is concerned with the fabrication of a composite material consisting of a two-phase ceramic skeleton, i.e. boron carbide and in situ generated titanium diboride with excess free carbon, reaction bonded to silicon carbide that was formed subsequently to the infiltration of the skeleton with molten silicon. The material displays high rigidity, extremely high hardness and low density, providing potential for applications ranging from light armor to precision machining equipment. Its advantage resides in the significantly lower sintering temperature than is necessary for pressureless sintering of boron carbide. The advantage of this approach resides in the several degrees of freedom that allow some latitude in designing required properties.

INTRODUCTION AND BACKGROUND

Hard ceramic materials are playing an ever increasing role in armor applications. Light ceramics are particularly attractive for personnel, land and airborne vehicle protection. Elevated hardness values are by common consensus of crucial importance for good ballistic resistance. Thus, it is not surprising that boron carbide, the hardness of which is second only to that of diamond and which has one of the lowest mass per volume ratio has long been considered as a choice armor material candidate. In addition to the above mentioned properties, ease of fabrication and costs are of paramount importance in determining whether a particular material is considered for actual armor applications.

Monolithic boron carbide of full density can be fabricated only by hot pressing at elevated temperature. Near-full density, a pre-requisite for achieving acceptable ballistic resistance, can be achieved in boron carbide only in the presence of sintering additives, the most common of which is free carbon. The intrinsic limitations of the hot pressing technology and the elevated equipment costs have limited the widespread use of boron carbide for personnel armor, even though it is by far superior to the commonly employed light armor materials, namely, aluminum oxide.

A break-through occurred in the late nineties with the development by several US companies of the "reaction bonding" process for the fabrication of SiC (RBSC) or/and B_4C (RBBC) armor material. The significantly reduced production costs that ensue from the lower processing temperatures, of the order of 1600°C instead of 2200°C - that is used for the fabrication of monolithic boron carbide - revolutionized the light armor production scene. With the advent of the Iraq war, the US department of defense placed orders for tens of thousands of Small Arms Protective Insert (SAPI) units.

Essentially, according to the reaction bonding process, unsintered powders of SiC, B_4C or their mixture with the addition of free carbon are compacted and subsequently infiltrated with molten silicon. Molten carbon reacts with free carbon and possibly with some carbon from within the boron carbide to form a continuous SiC matrix in which the original B_4C and/or SiC particles are embedded, along with some pockets of residual unreacted SiC. The resulting composite has a very high cohesive strength, displays elevated values of hardness and has been proven to be an effective ballistic impact resistant material.

Numerous papers that appeared in scientific journals as well as patents describe several variants of reaction bonding processing as well as the properties of the final composite.[1,2,3,4] One of the drawbacks associated with the fabrication of reaction bonded boron carbide (RBBC) is the requirement for the presence of excess free carbon. In its absence, molten silicon reacts with carbon extracted from the boron carbide, decreasing the strength of the latter. Moreover, the lack of carbon may lead to the increased formation of free silicon pockets which are weak spots in the composite material. Carbon is added either in the form of pure carbon or in the form of carbon-rich organics. The latter carbon-containing compounds have to be pyrolized leaving behind free carbon that is dispersed throughout the compacted boron carbide in various forms. Pyrolisis is an environment unfriendly procedure, moreover, in many cases, residual free carbon instead of being dispersed uniformly forms agglomerates that react partly only with the molten silicon and detract from the final strength of the composite material.

Excess free carbon and its uniform dispersion arose as an issue also in connection with the pressure-less sintering of boron carbide well before the application of the reaction bonding technique via infiltration with molten silicon. Sintering of boron carbide takes place only at temperatures close to the melting point and is associated with excessive grain growth. The presence of free carbon as an additive was found to assist sintering and numerous efforts were made to develop and optimize the mixing and dispersion parameters with the boron carbide powder. Sigl et al.[4,5] suggested the addition of titanium containing compounds that would react with boron carbide resulting in the formation of TiB_2 and in the in-situ release of free carbon. The released free carbon was supposed to act as a sintering enhancing additive and reduce the temperature necessary for pressureless, full densification of a two-phase, $B_4C + TiB_2$, ceramic. According to Sigl et al.[4,5] these expectations were borne out and high quality nearly fully dense ceramics were obtained.

Returning to the reaction bonding process, it seemed that the above described approach could also be applied to ensure the presence of a controllable amount of free carbon dispersed uniformly on a microscopic scale in a boron carbide powder to be infiltrated by molten silicon. Titanium carbide was an obvious choice as additive since its reaction with boron carbide proceeds according:

$$B_4C+2TiC=2TiB_2+3C \qquad (1)$$

with free carbon being released ubiquitously, in-situ throughout the mixture of the initial carbide powders. The objective of this approach is not to ensure achieving full density but to provide a source of uniformly dispersed free carbon for the reaction bonding with silicon.

The advantage of the RBBC approach as compared to conventional sintering or hot pressing of monolithic boron carbide for the manufacture of hard and impact resisting ceramics resides in the skipping of the high temperature stage above 2200°C. According to the reaction bonding approach, infiltration of the compacted but unsintered ceramic powder mixture takes place below 1600°C, a significantly lower temperature that is associated with lower manufacturing costs. In the resulting material, the initial ceramic powder particles, B_4C, SiC or a mixture of thereof, are embedded and bonded by the newly formed SiC that forms a continuous matrix. The RBBC does display excellent properties as reported by Aghajanian et al.[2]

In principle, the properties of this boron carbide-based composite could be further improved by forming, prior to infiltration a continuously bonded ceramic skeleton. In other words, the compacted ceramic powder has to undergo a sintering treatment at a temperature lower than that required for obtaining full density, thus maintaining some of the advantage of the reaction bonding process. Reaction (1) takes place in the course of the treatment resulting in a two-phase preform with an interconnected network of pores and well distributed free carbon. The porous preform is subsequently infiltrated by molten silicon which reacts with excess carbon to form SiC, a third hard ceramic phase.

The present communication is concerned with the details of this approach and its anticipated advantages. These reactions are foremost the lower sintering temperature and the additional degrees of freedom for the reaction bonding process that pave the way for further optimization of the desired properties

EXPERIMENTAL PROCEDURES

The composites were prepared using relatively low cost B_4C powder (5μm mean particle size, Mudanjiang Abrasives And Grinding Tools Co, China) and TiC powder (Cerac®, -325 mesh). The powder mixtures (10, 20, 30 wt.% TiC) were homogenized in planetary ball mill for 2 h. The homogenized powders were uniaxially compacted at 15 MPa. Synthesis of B_4C-TiB_2-C multiphase ceramics from B_4C-TiC mixtures were conducted in a vacuum furnace (10^{-4} torr) at 1773 K for 4h. After the synthesis the compacts were sintered in the 2070-2130°C temperature range for 30 min in order to obtained the required open porosity, in the 10 to 30 vol.% range. The densities of the porous performs were determined by the liquid displacement method in distilled water. The porous preforms were infiltrated with liquid Si in a vacuum furnace (10^{-4} torr) at 1480°C for 20 min. The microstructure of the infiltrated composites was examined by Optical Microscopy and SEM (JEOL®JSM-5600). Phase composition was determined by X-Ray diffraction. (Rigaku RINT 2100). Three point bending test were conducted using 3x4x20mm bars. Vickers hardness (Buehler - Micromet 2100) was measured with a 2000 gram load. The elastic moduli of the composites were derived from ultrasonic sound velocity measurements using 10 mm diameter samples with plane parallel polished surface to within 0.005 mm.

RESULTS AND DISCUSSION

The thermodynamic analysis of the B-Ti-C ternary system in conjunction with an in-situ reaction of boron carbide with TiC that generates elemental carbon and TiB_2, was performed by Sigl et al.[4,5] The 2000°C isothermal section is shown in Fig.1 and indicates that the equilibrium state of a mixture of B_4C and TiC, in which the latter is a minority constituent, is located in the three-phase $B_4C + TiB_2 + C$ region. Sigl et al.[4,5], also reported that the addition of TiC to B_4C improves the sintering behavior of the mixture.

In principle the porosity of the two-phase $B_4C + TiB_2$ preform depends on several parameters. The processing parameters, i.e. sintering temperature and its duration exert the major influence; the TiC-to-B_4C initial ratio also affects the final result, albeit to lesser extent (Fig.2).

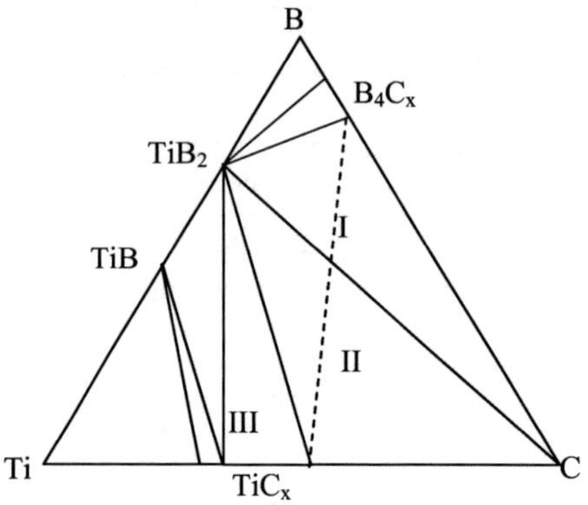

Figure 1. Ternary Ti-B-C diagram. The dashed line connecting TiC to B4C is the locus of all the phase compositions corresponding to different initial TiC-to-B_4C ratios

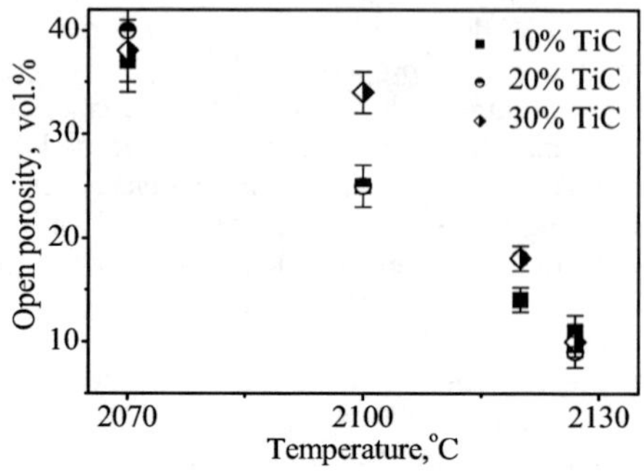

Figure 2. The dependence of the porosity of the preform on the sintering temperature for duration of 60 min and for various relative amounts of TiC in the mixture.

The initial porosity dictates the amount of Si that can be introduced into the preform and thereby affects the amount of SiC formed and determines the density of the infiltrated composites (Fig.3).

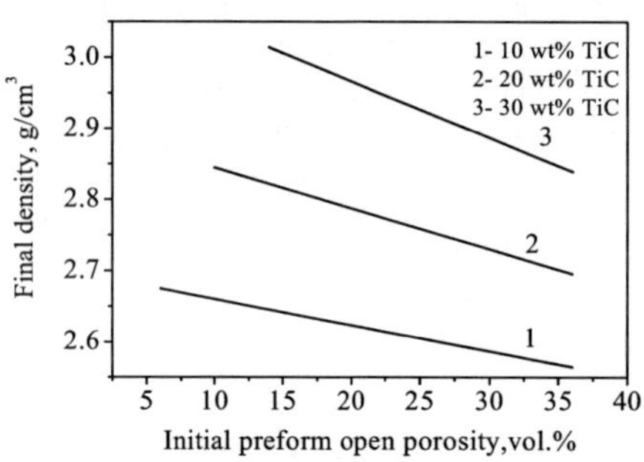

Figure 3. Calculated densities of the composites as a function of the initial porosity of the preforms and for various fractions of TiC.

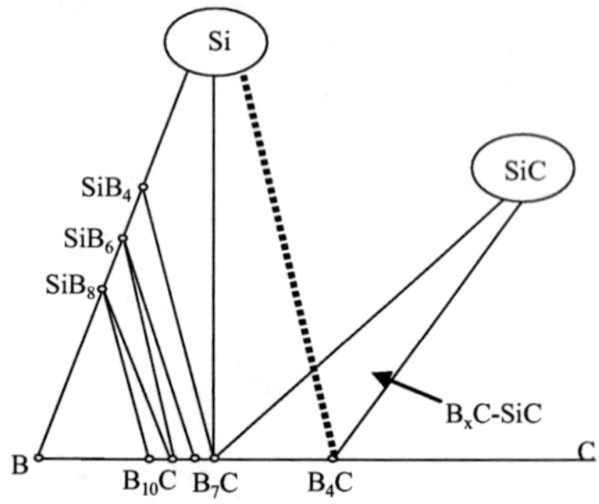

Figure 4. The boron rich corner of the ternary B-Si-C system.

The thermodynamic analysis of the B_4C + Si system has been carried previously in conjunction with wetting studies of B_4C substrates by various Si containing molten alloys and had been detailed elsewhere[6]. We recall that boron carbide has a relatively wide composition range of stability that extends from B_4C to $B_{10}C$. Thus, in principle, boron carbide may also become a source of carbon that may react with the infiltrating molten Si. The results of the analysis have shown that SiC and liquid silicon are in equilibrium with boron carbide at a composition that corresponds to formula B_7C (Fig.4)

Thus we expect to encounter at the outcome of the infiltration into the two-phase B_4C + TiB_2 preform, some carbon depleted boron carbide along with newly formed SiC. These predictions were fully confirmed by x-ray analysis WDS and EDS analysis. The phase composition of the sintered preforms and the infiltrated composites, according to X-ray and image analysis, is shown in Fig. 5 for two values of the open porosity of the preforms and for an initial B_4C-20wt.%TiC mixture.

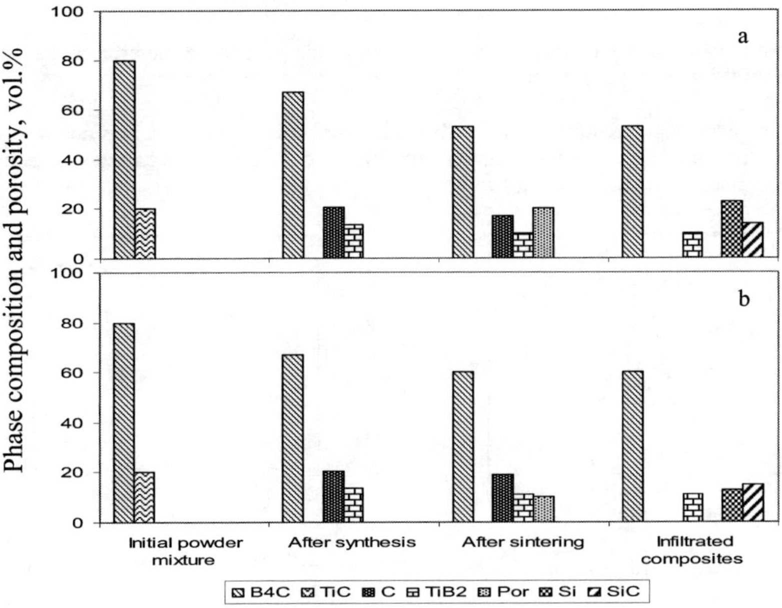

Figure 5. Phase composition of the preforms and infiltrated composites, a) the preform sintered at 2080°C with initial porosity of 20vol% and displaying after infiltration a 2200HV hardness value; b) the preform sintered at 2130°C with initial porosity of 10 vol.% and displaying after infiltration 3100HV hardness value.

Mechanical properties.

The elastic moduli of the multi-phase composite are shown in Fig. 6. These compare favorably with the moduli of the RBBC as reported by Aghajanian et al [2]. As expected the moduli decrease monotonically with increasing initial porosity that is associated with an increasing fraction of residual silicon.

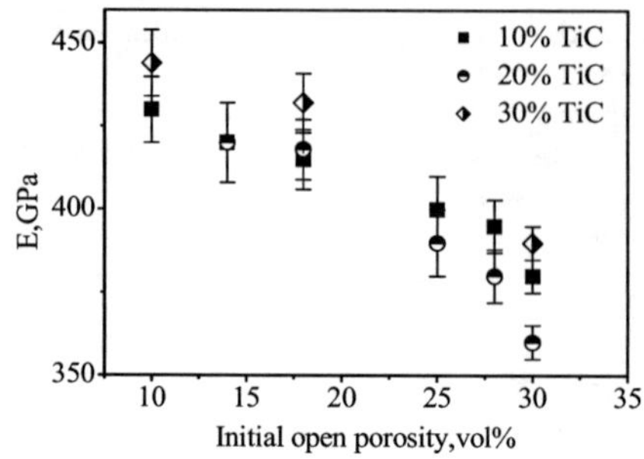

Figure 6. Elastic moduli of the composites as a function of the porosity of the preform and for various values of the initial TiC fraction.

The Vickers hardness (Fig.7) displays a similar behavior. In that case, however, it is noteworthy that the higher the TiC fraction in the initial powder mixture, the higher the hardness at equal initial porosity values.

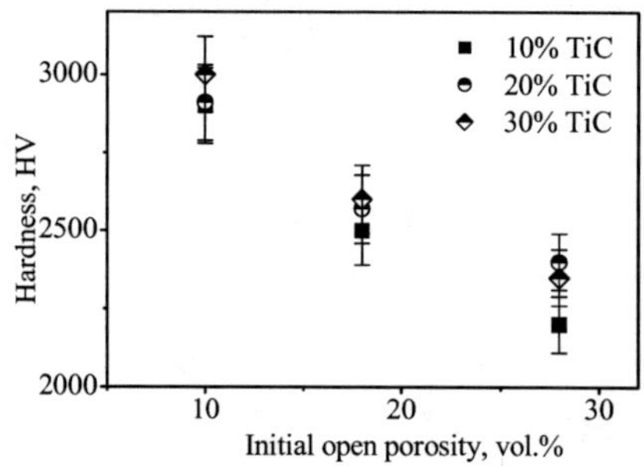

Figure 7. Vickers hardness of the composites as a function of the porosity of the preformand for various values of the initial TiC fraction.

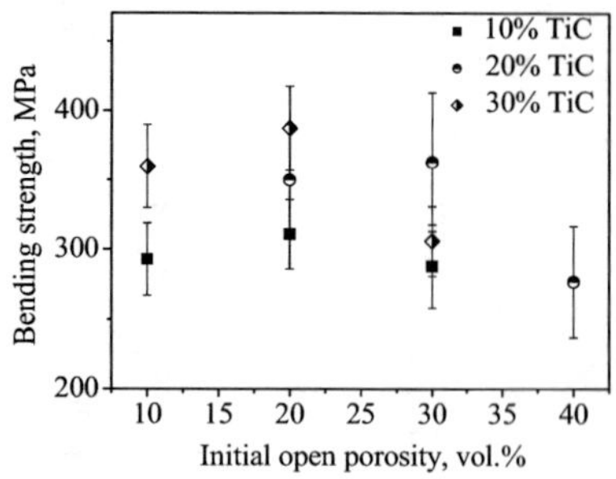

Figure 8. Flexural strength of the composites as a function of the porosity of the preformand for various values of the initial TiC fraction.

The flexural strength values of the composite are shown in Fig (8). Whereas it is not possible to detect a clear dependence on initial porosity, the results indicate that the initial TiC fraction exerts a significant effect. For all initial porosity values, the increased relative TiC fraction in the initial mixture, at equal porosity, increases the flexural strength by about 20 pct.

Microstructure

The scanning electron fracture surface microstructure of the sintered yet non-infiltrated ceramic preform is shown in Fig. (9). This porous perform after the early stages of the sintering consists of two ceramic phases, B_4C and TiB_2 and some excess carbon resulting from the reaction (1) that took place at approximately 1500°C. The interconnected network of pores with channels of the order of 10 μm diameter is commensurate with the initial particle size of the powder particles. The TiB_2 particles which appear brighter in the backscattered image, due to their higher atomic molecular weight are well fused to the initial B_4C particles and form with the latter a robustly interconnected solid network.

Figure 9. Scanning electron micrograph of the fracture surface of the ceramic prefoms consisting of B_4C, TiB_2 and excess carbon

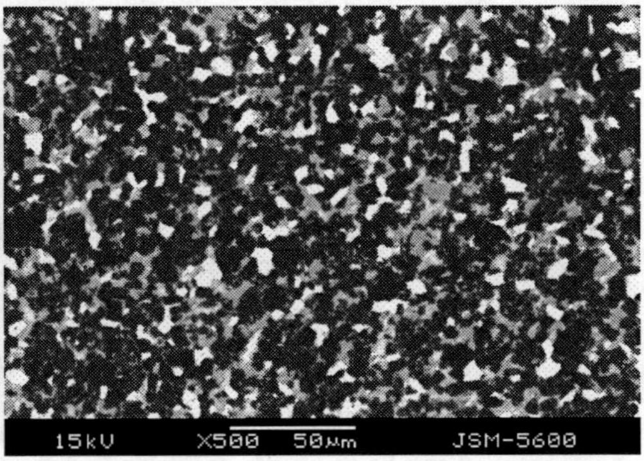

Figure 10. The relatively low magnification of this micrograph does not allow distinguishing between the SiC and the free Si areas which appear in grey in the picture.

The uniform distribution of the various ceramic phases is well illustrated in the backscatter electron micrograph shown in Fig. 10

The SEM image of the infiltrated composite is shown in Fig. (11). Four different phases can be distinguished. The majority phase which appears darkest in the micrograph is the original B_4C, part of the ceramic network, while the brightest is the TiB_2 phase, product of the reaction between the former and the original TiC.

The particles of the two ceramic phases have in general well rounded smooth grain boundaries as the result of the high temperature sintering treatment. The two intermediate colored gray phases, SiC, dark grey and Si, lighter grey, were formed after the infiltration. Silicon carbide was formed as the result of the reaction between Si and excess carbon. In reaction bonded boron or silicon carbide, pockets of unreacted silicon are always present. These pockets detract

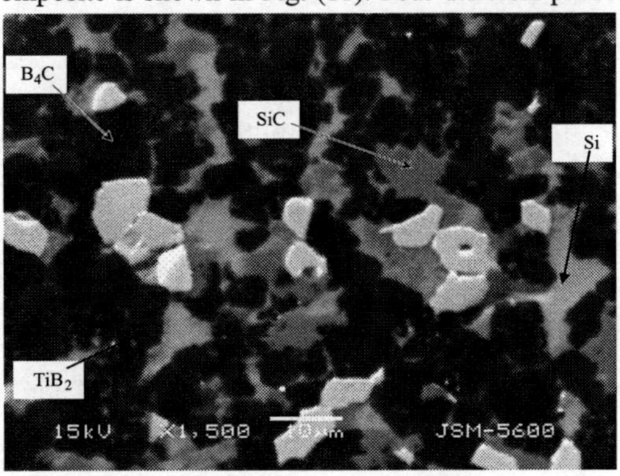

Figure 11. Backscatter micrograph of the infiltrated composite. No residual porosity is visible. Notice the relatively uniform grain size of the different phases

from the strength of the composite but enhance its toughness. The sharp edges of both the Si and SiC grains reflect the rapid solidification of a liquid phase and its rapid reaction with the free carbon dispersed throughout the ceramic preform.

SUMMARY AND CONCLUSIONS

A processing approach for the production of a very hard and stiff composite that may be useful as a light armor material is described. An initial mixture of B_4C and TiC heated to 1500°C provides in-situ generated free carbon distributed homogeneously throughout a two-phase ceramic preform consisting of B_4C and TiB_2. Further heating to within the 2050 - 2150°C temperature range allows retaining 10 to 30% residual porosity.

Molten silicon infiltrates the porous preform at 1500°C, reacts with the free carbon to produce SiC, a third, hard ceramic phase, reaction bonded to the two ceramic phase skeleton. In addition, some 5 to 7% residual Si is retained in the ceramic composite.

The hardness of the infiltrated composites lies between 2700-3000 HV, the elastic moduli are in the 400-450 GPa range and the bending strength is about 320-380 MPa. One of the main advantages of this processing approach is that it provides several degrees of freedom that pave the way for further optimization of the static and the dynamic mechanical properties.

ACKNOWLEDGMENT

This work was partly supported by the Israel Science Foundation grant 118/03.

REFERENCES

[1]K.M. Taylor and RJ, Palicke, "Dense Carbide Composite for Armor and Abrasives", U.S. Pat. No. 3 765 300 Oct. 16, 1973.

[2]M.K. Aghajanian, B.N. Morgan, J.R.Singh, "A new family of reaction bonded ceramics for armor applications", in "*Ceramic Armor Materials by Design*", Eds. J.W.McCauley et al., Ceramic Transactions, **134**, 527-544 (2002)

[3]I.S. Han, K.S. Lee, D.W. Seo, S.K. Woo, "Improvement of mechanical properties in RBSC by boron carbide addition" , J. Mater. Sci. Lett. **21**, 703– 706 (2002).

[4]L.S.Sigl, "Processing and Mechanical Properties of Boron Carbide Sintered with TiC", *J.Europ. Cer.Soc.*, **18**, 1521-29 (1998);

[5]L.Sigl, H.Thaler, K-A. Schwetz, U.S. Patent No. 5543370 (1994).

[6]Frage, N.; Froumin, N.; Aizenshtein, M.; Dariel, M.P., "Interface reaction in the B4C/(Cu-Si) system", *Acta Materialia*, **52**, 2625-2635 (2004)

BIOMORPHIC REACTION BONDED SILICON CARBIDE CERAMICS FOR ARMOR APPLICATIONS

Bernhard Heidenreich and Michaela Gahr
Institute of Structures and Design
German Aerospace Center
Pfaffenwaldring 38-40
70569 Stuttgart
Germany

Eugene Medvedovski
UIP
Providence, RI, USA
formerly with Ceramic
Protection Corporation
Calgary, Canada

ABSTRACT

Reaction bonded silicon carbide (RBSC) ceramics demonstrate high performance at ballistic protection applications. Biomorphic SiSiC ceramics based on wooden preforms and manufactured via pyrolysis and liquid silicon infiltration (LSI) are considered as a promising armor material due to their remarkable level of physical properties and a cost benefit. Manufacturing process, structure and physical properties of the biomorphic RBSC and their fracture behavior under ballistic impacts have been studied. The armor systems based on the biomorphic RBSC ceramic tiles of various thicknesses can defeat 7.62x51-mm NATO Ball FMJ, 7.62x54R LPS, 7.62x63-mm AP M2 ammunitions.

INTRODUCTION

Ceramic armor systems consist, in general, of a monolithic ceramic body bonded with a soft but high tensile strength backing material such as special fiber lining (e.g. Kevlar™, Twaron™, Spectra™ or fiberglass) and, sometimes, with soft metals (e.g. aluminum). Upon impact of the ballistic projectiles with a high velocity (700-900 m/sec or greater) and kinetic energy of 2-4 kJ, the hard-faced ceramic is cracked and broken, and the residual energy is absorbed by the soft reinforced backing material. This backing material also supports post-impact fracturing of the ceramic body and the defeated bullet. Structure and properties of a ceramic facing material and its manufacturing features are significant factors affecting ballistic energy dissipation and, hence, performance of ballistic protection systems.

Among different advanced armor ceramics, non-oxide materials such as boron carbide (B_4C), silicon carbide (SiC), silicon nitride (Si_3N_4), aluminum nitride (AlN) and some others, including the materials based on their binary systems, have high physical properties and relatively low density that is beneficial for ballistic protection applications [1-8]. However, these ceramics are often manufactured by hot pressing that is relatively expensive and not very productive. Although pressureless sintered materials, such as commercially produced SiC ceramics, are less expensive than hot-pressed materials, they are still relatively expensive because their manufacturing requires kilns with special controlled atmospheres and very high temperatures for sintering. Dense homogeneous boron carbide and silicon carbide ceramics have low fracture toughness and high brittleness, and therefore, they do not have desirable multi-hit ballistic performance. Reaction-bonded silicon carbide (RBSC) and some other reaction-bonded carbide-based ceramics are considered as prospective materials for armor applications due to relatively lower cost than hot pressed or pressureless sintered ceramics, high physical properties and an ability to manufacture relatively large sized armor products [9, 10]. Due to their heterogeneous structures and associated lower brittleness, these ceramics demonstrate much better integrity for multi-hit situations than dense homogeneous carbide-based ceramics.

Ceramic-matrix composites also demonstrate a high integrity after ballistic impact due to their mechanical properties and impact energy dissipation ability. The following ceramic-matrix composites are mentioned as armor materials [3]: ceramic reinforced with whiskers or fibers, such as compositions of Al_2O_3/SiC_w, Al_2O_3/SiC_f or Al_2O_3/C_f, and ceramics/particulate-based compositions (TiB_2/B_4C_p, TiB_2/SiC_p). Cermets such as Lanxide™ composites based on SiC infiltrated with Al, Ni/TiC, Al/B_4C_p and some others also demonstrate superior qualities. The majority of these materials are hot-pressed, and, therefore, expensive. Although some metal-infiltrated composites, such as Lanxide™ SiC/Al composite, or SiC/SiC composites consisted of SiC matrix reinforced with SiC fibers [8], are not hot-pressed, they need special processes and equipment; they are also relatively expensive and are prone to difficult problems in manufacturing.

RBSC ceramics, which also may be considered as ceramic-matrix composites, may be manufactured using several different approaches; they include a silicon melt infiltration into a porous preform, consisting of carbon or silicon carbide and carbon structures, in vacuum or in an inert atmosphere. For example, RBSC ceramics may be produced via a molten silicon infiltration of low-modulus carbon-fiber/carbon preforms [11]. Another approach includes a molten silicon infiltration of the porous preforms from SiC grains with a specially selected particle size distribution and carbon, which can be prepared by casting or pressing (as a carbon source, either carbon or pyrolysed organic binders may be utilized). This "conventional" approach is presently successfully using by several companies for ceramic armor manufacturing [9, 10]; one of such materials is described by M. Aghajanian et al. [9]. In the present work, relatively novel reaction-bonded ceramic-matrix composite materials based on liquid silicon infiltration (LSI) of pyrolysed wooden templates have been considered as armor materials. Such biomorphic RBSC ceramics provide excellent capability to manufacture complex and large shapes, lower cost and reproducible shrinkage [12, 13]. It was shown [14] that biomorphic RBSC materials have some potential for armor applications opposite to fiber-reinforced C/C-SiC composites also prepared via LSI. In the present work, manufacturing process, structure, physical properties and ballistic performance of the biomorphic RBSC ceramics have been studied.

EXPERIMENTAL

Materials and Manufacturing

Biomorphic RBSC ceramics were manufactured using commercially available medium dense fiber (MDF) boards. These panels are widely used in the furniture industry and are made by pressing from fine fibers of needle wood with phenolic resin in a mass production process. MDF preforms usually have bulk density of 0.6-0.9 g/cm^3. Silicon granules were used for infiltration and siliconization of the pyrolysed wooden preforms.

Wooden boards were dried at 100-110°C and then pyrolysed at temperatures up to 1650°C in inert atmosphere. A relatively high temperature of pyrolysis is selected due to the fact that, at this temperature level, a complete outgassing in the preform occurs, and, as a result, the silicon melt does not contain the gaseous phase that provides a higher level of infiltration and lower closed porosity of the final material [15]. The prepared preforms were infiltrated with silicon at 1600-1700°C under vacuum. All parameters of the heat treatment processes, including siliconization (e.g. heating rates, vacuum level, etc.), were optimized. Initial dimensions of the wooden boards were selected based on the consideration of a linear shrinkage during pyrolysis of approximately 25%. The ceramics practically do not shrink during siliconization process. After cooling, the tiles were ground with diamond tools. Ceramic tiles used for ballistic testing had a format of 100x100 mm and a thickness of 9.5, 7.5 and 6 mm.

Samples for ballistic testing were built up as two layers targets consisting of ceramic tiles bonded with aramid-based backing materials (e.g. Kevlar™ or Twaron™) widely used for ballistic protection. Ceramic tiles wrapped with fiberglass (prepreg) were bonded with proper amounts of aramid layers. The backing material type and thickness were selected on the experimental basis depending on the thickness of ceramics and on the ammunition used for ballistic testing. The bonding was conducted in autoclave (i.e. at elevated temperature and pressure) using earlier developed procedure. Due to a small format of ceramic tiles and possible related excessive fragmentation upon ballistic impact, the target tiles were surrounded with alumina tiles or bonded as small panels of 2x2 tiles.

Testing

Microstructure was studied using scanning electron microscopy (SEM) for fracture surfaces and optical microscope for polished sections. X-ray diffraction (XRD) analysis was conducted for the final (siliconized) sample. Density was tested using the water immersion method based on the Archimedes law. Sonic velocity and Young's modulus were tested by the ultrasonic technique measuring the longitudinal ultrasonic velocity in accordance with ASTM C769 and by the resonant frequency method in accordance with ASTM C885. The formula $E=V_l^2\rho(1+\upsilon)(1-2\upsilon)/(1-\upsilon)$ was used for the calculation, where E is the Young's modulus, V_l is the longitudinal sonic velocity measured in accordance with ASTM C769, ρ is the density, υ is the Poisson's ratio (Poisson's ratio was considered as 0.11-0.17 based on the ASTM C885 test results for various SiC-based ceramics). Three-point flexural strength was tested in accordance with a procedure similar to ASTM C113. Knoop and Vickers hardness were tested in accordance with ASTM C1326 and C1327, respectively, at the indentation load of 1 kg. Fracture toughness (critical stress intensity factor) K_{Ic} was determined using the indentation technique based on the samples prepared for Vickers hardness testing and using the same load; K_{Ic} was calculated using the well-known formula: $K_{Ic}=0.941Pc^{-3/2}$, where P is the indentation load and c is the crack length measured under microscope. The test samples with required dimensions were cut from the mentioned test tiles.

The ballistic performance of the ceramics bonded with appropriate backing materials was tested using weapons such as the M16, AK47 and some others (caliber 0.30). Depending on the application and the required level of protection, the ammunition 7.62x51-mm NATO Ball Full Metal Jacket (FMJ) with a lead core, 7.62x54R LPS with a steel core and 7.62x63-mm armor-piercing AP M2 with a tungsten carbide core were used. Depending on the ammunition, the bullet weight, velocity and energy were varied, e.g. the bullet velocities for the mentioned projectiles were 830-870, 700-730 and 820-

870 m/s, respectively. The bullet velocity during testing was measured using an optical chronograph. The trauma after shooting was evaluated using a Roma Plastilina modeling clay placed behind the armor system; the trauma in clay shows the transient deformation of the composite on the back of the system. The damage zone of the ceramics, including ceramic fragmentation, and the subsequent post-impact condition of the bullet, were observed.

RESULTS AND DISCUSSION

It is known that a rapid silicon infiltration and a complete SiC formation occur if the preforms have large enough pore sizes. The pyrolysed wooden preforms from MDF boards used in the present work have hollow channels with cells of various sizes ranged from several μm to 100 μm that is acceptable for silicon infiltration and for the SiC formation without stresses, which may be caused by a volume increase during this SiC formation. The prepared biomorphic RBSC ceramics consist of silicon carbide (major phase) formed due to a high-temperature reaction of carbon derived from pyrolysis of the wooden preform with molten silicon and residual silicon. The ratio between silicon carbide and silicon is approximately 60-65 vol.% and 30-35 vol.%, respectively; a very small amount of carbon (up to 3 vol.%) is also observed. This ratio was estimated in accordance with the density calculation method described by Q. Guanjun et al. [16]. In general, in biomorphic RBSC ceramics, the structure of the wooden precursor is preserved. The structure of the ceramics is rather homogeneous (Fig. 1), i.e. more uniform than different RBSC ceramics made by the siliconization of SiC-C preforms.

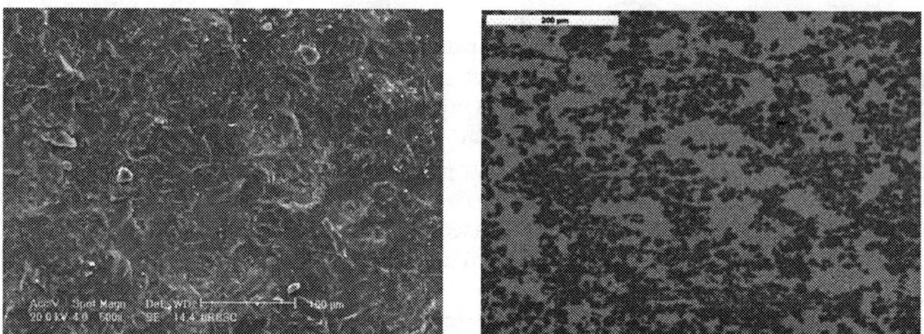

Figure 1: SEM image of biomorphic RBSC ceramics (left: 500x; right: 200x, polished, cross section).

Silicon is distributed rather uniformly between the SiC grains; however, some areas with relatively large (5-15 μm) and uneven silicon distribution are observed. Silicon carbide grains have an average size of 5-20 μm; however, grain boundaries are not clear. XRD analysis indicated mostly β-SiC that correlates well with literature data [13]. Due to a relatively high content of residual silicon, density of the studied biomorphic RBSC ceramics is 2.79-2.81 g/cm^3 (water absorption is not greater than 0.02%) that is lower than of "conventional" armor RBSC ceramics formed by the infiltration of coarse-grained SiC-C preforms with molten silicon (2.98-3.07 g/cm^3), i.e. these ceramics provide approximately 7-8% lower weight.

Mechanical properties of the biomorphic RBSC ceramics (see Table 1) are defined by their phase composition and structure, but they are on the remarkable level. Generally, a major phase of these ceramics has hardness lower than of the "conventional" RBSC ceramics. For example, HK1 and HV1 values of biomorphic RBSC ceramics are 1600-1700 kg/mm^2 and 2000-2100 kg/mm^2, respectively, vs. 2000-2150 kg/mm^2 and 2350-2450 kg/mm^2 for RBSC ceramics. Due to a relatively homogeneous structure with a relatively thin area of residual silicon, hardness for biomorphic RBSC ceramics was tested only for the major phase, and the influence of "bonding" between grains could not be studied. Fracture toughness K_{Ic} of biomorphic RBSC ceramics is also lower than of "conventional" RBSC ceramics. Similar situation is observed for the comparison of Young's modulus, specific stiffness (Young's modulus/density) and sonic velocity of these two types of RBSC ceramics. It should be noted that lower values of Young's modulus for biomorphic RBSC ceramics are also dealt with lower values of ceramic density although their sonic velocity values are very satisfactory for armor ceramics (greater than 10,000 m/s). However, flexural strength of the studied biomorphic RBSC ceramics can be compared with strength of "conventional" RBSC ceramics due to more homogeneous and fine-grained structure.

Table 1: Physical Properties of the Studied Biomorphic RBSC Ceramics in Comparison with "Conventional" Armor RBSC Ceramics

Property	Biomorphic RBSC	"Conventional" RBSC***
Density, g/cm^3	2.8	3.0-3.07
Sonic velocity, km/s	10.4	10.3-11.6
Young's modulus, GPa	290	300-400
Specific stiffness, MPa.m^3/kg	105	100-130
Knoop hardness* HK1, kg/mm^2	1650	2000-2150
Vickers hardness* HV1, kg/mm^2	2050	2350-2450
Fracture toughness K_{Ic}, MPa.m$^{0.5}$	2.3	2.2-2.8
Flexural strength**, MPa	230	190-250

* tested for the major phase

** biomorphic RBSC ceramics tested using 3-point test while the data for "conventional" RBSC ceramics accumulated for 4-point test

*** the data accumulated for different RBSC ceramics manufactured by several different producers

The biomorphic RBSC ceramics demonstrated satisfactory ballistic performance if the tiles were bonded with properly selected backing material systems. It was found that, depending on a thickness of the tiles, a proper type of aramid backing as well as a thickness of the backing (amount of plies) should be selected. The designed armor systems based on the studied ceramics defeated 7.62x51-mm NATO Ball FMJ and 7.62x54R LPS (tiles with a thickness of 6 and 7.5 mm) and even 7.62x63-mm AP M2 (tiles with a thickness of 9.5 mm). The trauma in clay was on the acceptable level similar to other types of armor ceramic tiles with the same dimensions. Results of the ballistic tests are shown on the Fig. 2 - 4. It should be noted that the achieved results are similar to the ballistic performance of the "conventional" RBSC armor ceramics in terms of the thickness of ceramic tiles and armor system designs. Satisfactory ballistic performance of the system with biomorphic RBSC against AP projectiles with WC cores was achieved although generally RBSC ceramics do not demonstrate high ballistic performance in this situation due to possible transformation of silicon under ballistic impacts (phase transformation of silicon under mechanical loads was noted by V. Domnich and Y. Gogotsi [17]).

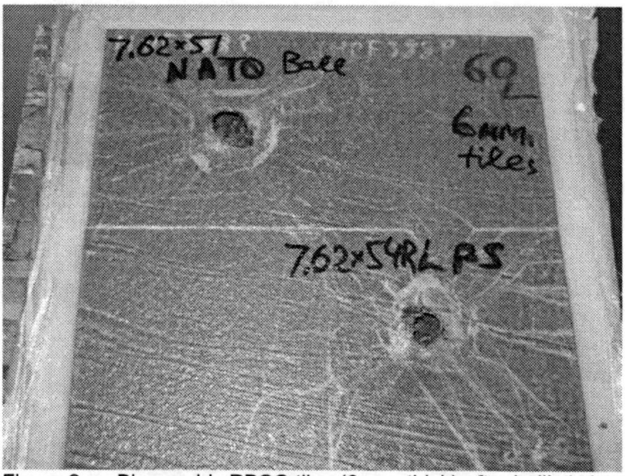

Figure 2: Biomorphic RBSC tiles (6 mm thick) after ballistic impact with 7.62x51-mm NATO Ball FMJ and 7.62x54R LPS ammunitions, no penetration of the system

Figure 3: Biomorphic RBSC tile (9 mm thick) after ballistic impact with 7.62x63-mm AP M2, the backing was taken off. It is clear seen that there is no penetration of the backing after shooting. The details are shown in Figure 4.

Figure 4: Biomorphic RBSC tile (9 mm thickness) after ballistic impact of 7.62x63 mm AP M2 (left). Fracturing of biomorphic RBSC ceramics after ballistic impact (right).

Upon ballistic impact, the ceramic targets were fracturing resulting in formation of the chunks and rather fine powder. It should be noted that the powder formation was greater than in the case of "conventional" armor RBSC ceramics with heterogeneous structures, but much less than in the case of dense homogeneous pressureless sintered SiC ceramics. The conoidal cracks were mostly observed, although some other types of cracks were also found but not so clear due to elevated "powdering". As thinner ceramic tile was ballistically shot, as more powder formation was observed. The noted fracturing behavior is dealt with phase composition and microstructure of the considered biomorphic material, i.e. with an elevated content of residual silicon and a fine-grained microstructure. However, a two-phase composition of the studied ceramics promotes a higher energy dissipation ability and lower shattering upon ballistic impact than dense homogeneous SiC ceramics (hot pressed or pressureless sintered) that makes this type of RBSC ceramics promising for ballistic protection, including for multi-hit applications. It may be assumed that fracturing and crack propagation occur through the silicon phase due to its significantly lower mechanical properties comparatively with silicon carbide.

It should be noted that ballistic performance of biomorphic RBSC ceramics may be improved by optimization of their microstructure strongly dependant on the structure of starting MDF panels and pyrolysed preforms (e.g. pore size distribution that defined by the shape and size of wooden chips and a content of phenolic resin used for MDF panel fabrication). Based on the practical experience, there is a possibility that less homogeneous biomorphic ceramics with larger SiC grains but with the same or lower content and size of a residual Si phase may have very satisfactory ballistic performance, including in the multi-hit situations, although some mechanical properties may have lower values. Modification of the ceramic structure based on the proper selection of MDF preforms and ballistic testing of the prepared biomorphic ceramics require further studies.

Based on the positive results of ballistic tests conducted with different projectiles, the studied ceramics may be considered as very promising for armor applications, including for personnel, vehicular, aircraft and structural protection (e.g. blast protection). A high potential of these ceramics is also dealt with an ability to manufacture large size products. It is known that large armor panels consisting of ceramic tiles ("mosaic" design) have weaker areas near the joints of the tiles; in order to eliminate lower ballistic performance, the tiles are produced with a greater thickness or the tiles should have more complicated design (e.g. with a thicker area near joints) or this joints should be additionally protected, i.e. such armor systems have additional weight. The use of monolithic armor tiles without joints can eliminate the mentioned problem, and large-sized biomorphic RBSC ceramics that may be manufactured utilizing simple inexpensive preforms is a good candidate for such applications. Large size armor panels made from this type of ceramics should be ballistically tested to verify their ballistic performance.

CONCLUSIONS

Structure and physical properties of biomorphic RBSC ceramics prepared via pyrolysis of wooden preforms with following siliconization have been studied. Biomorphic RBSC ceramics may be considered as a prospective material for armor applications due to a remarkable level of mechanical properties, lower density with practically zero open porosity providing lower weight of armor and an ability to manufacture large and near net shape-products with a relatively low cost. Armor systems with optimized designs based on the tiles manufactured from the studied ceramics bonded with appropriate backing demonstrated satisfactory ballistic performance defeating 7.62x51-mm NATO Ball FMJ, 7.62x54R LPS, 7.62x63-mm AP M2 ammunitions. Biomorphic RBSC ceramics may be used as a component of armor systems for personnel ballistic protection, for vehicular and structural protection, e.g. as large monolithic panels.

REFERENCES

[1]. C.F. Cline, M.L. Wilkins, "The Importance of Material Properties in Ceramic Armor"; pp 13-18 in *DCIC Report 69-1*; Part I: "*Ceramic Armor*", 1969.

[2]. Soon-Kil Chung, "Fracture Characterization of Armor Ceramics", *American Ceramic Society Bulletin*, **69** [3] 358-66 (1990).

[3]. D.J. Viechnicki, M.J. Slavin, M.I. Kliman, "Development and Current Status of Armor Ceramics", *American Ceramic Society Bulletin*, **70** [6] 1035-39 (1991).

[4]. I.Yu. Kelina, Yu.I. Dobrinskii, "Efficiency of the Use of Silicon Nitride Ceramics as an Armor Material" (in Russian), *Refractories and Technical Ceramics*, [6] 9-12 (1997).

[5]. B. Matchen, "Application of Ceramics in Armor Products"; pp 333-342 in Key Engineering Materials, Vol. 122-124, *Advanced Ceramic Materials*. Edited by H. Mostaghasi. Trans. Tech. Publications, Switzerland, 1996.

[6]. R.G. O'Donnell, "An Investigation of the Fragmentation Behaviour of Impacted Ceramics", *Journal of Materials Science Letters*, **10**, 685-88 (1991).

[7]. V.C. Neshpor, G.P. Zaitsev, E.J. Dovgal, et al., "Armour Ceramics Ballistic Efficiency Evaluation"; pp 2395-401 in *Ceramics: Charting the Future*, Proceedings of the 8th CIMTEC (Florence, Italy, 28 June-4 July 1994). Edited by P. Vincenzini, Techna S.r.l., 1995.

[8]. T.M. Lillo, H.S. Chu, D.W. Beiley, et al., "Development of a Pressureless Sintered Silicon Carbide Monolith and Special-Shaped Silicon Carbide Whisker Reinforced Silicon Carbide Matrix Composite for Lightweight Armor Application"; pp. 49-58 in *Ceramic Armor and Armor Systems*, Ceramic Transactions, Vol.151. Edited by E. Medvedovski, 2003.

[9]. M.K. Aghajanian, B.N. Morgan, J.R. Singh, et al., "A New Family of Reaction Bonded Ceramics for Armor Applications"; pp 527-539 in *Ceramic Armor Materials by*

Design, Ceramic Transactions, Vol. 134. Edited by J.W. McCauley, A. Crowson, W.A. Gooch, Jr., et al., 2002.

[10]. E. Medvedovski, "Silicon Carbide-Based Ceramics for Ballistic Protection"; pp. 19-35 in *Ceramic Armor and Armor Systems*, Ceramic Transactions, Vol.151. Edited by E. Medvedovski, 2003.

[11]. K.L. Luthra, R.N. Singh, M.K. Brun, "Toughened Silcomp Composites – Process and Preliminary Properties"; *American Ceramic Society Bulletin* **72** [7] 79-85 (1993).

[12]. J. Schmidt et al., "Microstructure and Properties of Biomorphic SiSiC Ceramics Derived from Pyrolysed Wooden Templates"; pp. 414-419 in *High Temperature Ceramic Matrix Composites*. Edited by W. Krenkel et al., 2001.

[13]. Dong-Woo Chin, Sam Shik Park, Yong-Ho Choa, et al., Silicon/Silicon Carbide Composites Fabricated by Infiltration of a Silicon Melt into Charcoal; *Journal of the American Ceramic Society* **82** [11] 3251-3253 (1999).

[14]. B. Heidenreich, W. Krenkel, B. Lexow, "Development of CMC-Materials for Lightweight Armor"; pp. 375-381 in the Proceedings of the 27th International Cocoa Beach Conference on Advanced Ceramics and Composites: A. Vol. 24, Issue 3. Edited by W. Kriven and Hau-Tay Lin, 2003.

[15]. M. Gahr, J. Schmidt, W. Krenkel, et al., "SiC-Keramiken auf der Basis von Holzwerkstoffen"; pp. 383-388 in *Verbundwerkstoffe* (Ed.: H.P. Degischer), Wiley-VCH, Weinheim, Germany, 2003.

[16]. Q. Guanjun, M. Rong, C. Ning, et al., "Microstructure Transmissibility in Preparing SiC Ceramics from Natural Wood"; *Journal of Materials Processing Technology* 120, 107-110 (2002).

[17]. V. Domnich, Y. Gogotsi, "Phase Transformation in Silicon under Contact Loading"; *Reviews of Advanced Materials Science* **3**, 1-36 (2002).

REACTION-BONDED SiC COMPOSITES WITHOUT RESIDUAL Si

John P. Hurley,* Versha Singh, and John A. Hamling

University of North Dakota Energy & Environmental Research Center
PO Box 9018
Grand Forks, North Dakota 58202-9018
*Phone: (701) 777-5159, Fax: (701) 777-5181, E-mail: jhurley@undeerc.org

ABSTRACT

A method of creating near-net-shape silicon carbide (SiC) structures without residual silicon has been developed at the University of North Dakota Energy & Environmental Research Center (EERC). The technique involves slip-casting Si–SiC particle mixtures so that complex structures can be easily prepared, sintering the Si to the SiC, then converting the Si to continuous SiC. It does not include metal infiltration so no residual channels of Si remain in the structure, but it is 35% porous. The porosity reduces flexural strength but permits infiltration of a wide variety of polymers or ductile metals to make significantly tougher, continuous SiC composites with high residual compressive stresses. A preliminary ballistics test indicated that polymer-infiltrated SiC is much less prone to shattering than a dense SiC with similar stopping power. Metal-infiltrated SiC has not yet been tested. In addition, because the reaction bonding material begins as a slurry, it can be infiltrated into mats of woven high-temperature fibers to produce simple SiC–fiber composites. Production of these composites is only in the early stages.

INTRODUCTION

Because of its extreme hardness, SiC has the ability to shatter incoming projectiles during the dwell phase of impact.[1] Combined with its low density, it is a promising material for armor plating, although it also tends to shatter when struck by the projectile. The covalent molecular structure that makes the material so hard also makes it difficult to melt. Therefore, structures made of SiC are usually made through one of two methods, sintering or reaction bonding. In sintering, SiC powder is heated to temperatures over 1950°C[2] to cause the particles to bond together. To get the densest and strongest possible structures, pressure is applied at temperature. This requires relatively expensive processing equipment. Also, there is considerable shrinkage during sintering, and force can only be applied in the axial direction of the press so that only relatively simple shapes such as plates can be made by this process. In reaction bonding, a preform of porous C is prepared which may or may not contain SiC particles. Since C is easy to work, the preform can be made in complex shapes. Molten Si is then infiltrated into the structure to react with the C to form SiC, with Si metal filling the remaining pores. The resulting material is dense but is ultimately bonded together by Si metal as seen in its tendency to creep and corrode near the melting temperature of Si.[3–6] Although Si is a very hard metal, like the sintered SiC the reaction-bonded material is extremely brittle and tends to shatter when impacted.

Under a U.S. Department of Energy (DOE) program administered through the National Energy Technology Laboratory (NETL), the Energy & Environmental Research Center (EERC) has developed a new way of making reaction-bonded SiC which is not bonded with Si metal. It maintains the low-temperature processing and ability to make complex near-net shapes of other reaction-bonded techniques. However, it is composed solely of continuous SiC, so that although porous, it does not soften at temperatures over the melting point of Si. Therefore, it is not as strong as dense SiC, but it retains the ability to be infiltrated with polymers or more malleable metals to make composite structures with the hardness of SiC but possibly with greater resistance to shattering when impacted. By choosing the infiltrating medium appropriately, and coupling or decoupling the medium to the continuous SiC, it may be possible to dampen shock waves passing through the material through multiple reflections and refractions as the waves pass. In addition, by using an infiltrated medium that shrinks upon hardening, the structures can be placed into residual compressive stress. Much like prestressed concrete, the residual compressive stress can substantially increase the tensile strength of the material and decrease its likelihood of shattering on impact so that erosion of the projectile is more pronounced.[7] Finally, we are investigating the possibility of welding together individual segments of metal-infiltrated SiC to test if it could be used as a combined ballistic-protection and structural material. In this paper, we present an outline of the method of creating plates of the reaction-bonded materials along with micrographs of the structures and four-point flexure test data.

EXPERIMENTAL

The reaction-bonding material is made by mixing powdered Si and SiC with a binder and solvent to produce a thick slurry which is poured into a mold. The SiC is a green grade purchased from Washington Mills. The slurry is dried, and the binder is burned out either in air or by vaporizing it in inert gas. The plate is then heated in an inert atmosphere to sinter the material and bond the Si to the SiC. The Si in the sintered preform is converted to SiC by reacting it with condensed or gaseous C to form continuous, but porous, SiC. The pores can then be infiltrated with low-melting-temperature alloys or resins which are then polymerized.

RESULTS AND DISCUSSION

Figure 1 is a scanning electron microscope (SEM) picture of a polished cross section of a plate of sintered Si and SiC. The image was made with backscatter electron imagery in which brightness indicates the relative atomic number of the areas scanned. The white regions are the Si metal, gray regions are the SiC filler particles, and black regions are pores. This image shows that the Si is well bonded to the SiC in preparation for being converted to SiC. Four-point flexure tests show that the strength of these sintered plates is approximately 50 MPa. Generally, strength has been found to increase with the amount of Si that is used in the mixture, but shrinkage also increases, making it more difficult to produce a near-net-shape object. Typically, mixes are made so that shrinkage is less than 0.5%. Figure 2 shows a cross section of a plate in which the Si has been converted to SiC. The lack of white regions shows that the Si has been completely converted into SiC which bonds the initial SiC filler particles into a continuous microporous structure. It also indicates some redistribution of the Si during the carbiding step. Figure 3 shows the deflection of the flexure tester versus stress on the reaction-bonded material after sintering of the Si to the SiC and after the Si is converted to SiC. The data show that converting the Si to SiC doubles the strength to approximately 110 MPa while stiffness increases.

Figure 1. SEM picture of a polished cross section of a plate of sintered Si and Sic.

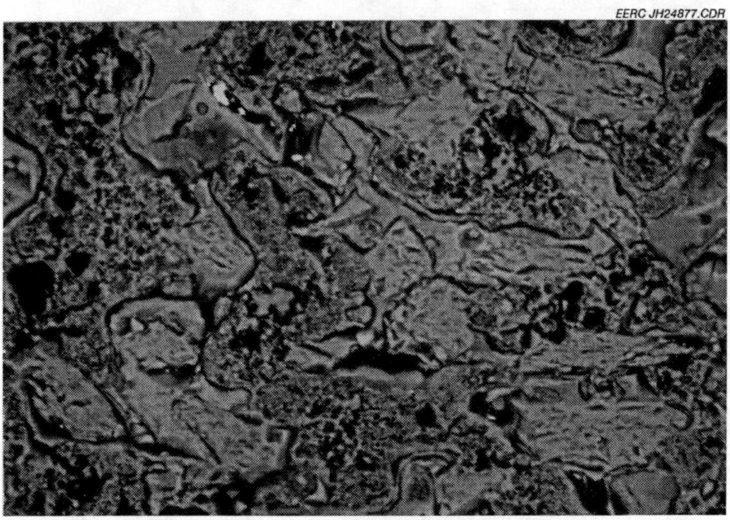

Figure 2. SEM picture of a polished cross section of a plate in which the Si has been converted to SiC.

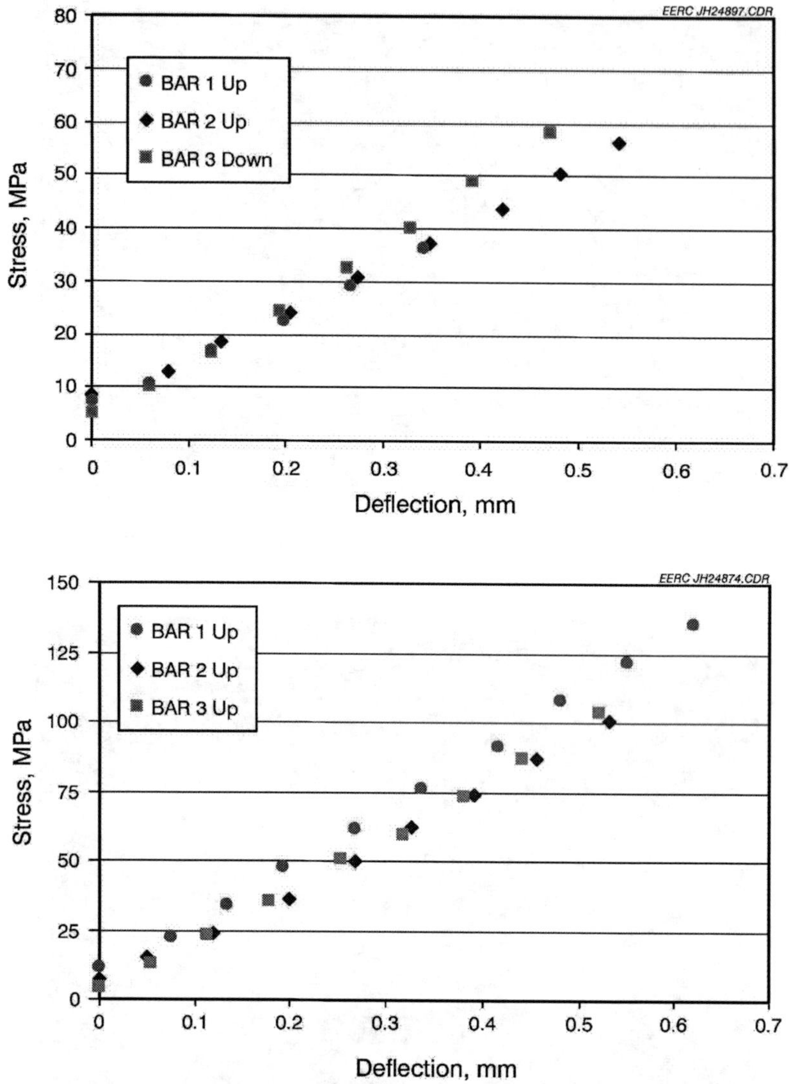

Figure 3. Deflection of the flexure tester versus stress of the material: (top) sintered Si/SiC and (bottom) Si converted to SiC.

Figure 4 shows the four-point flexure deflection versus stress data for the reaction-bonded SiC at 1450°C. It demonstrates that the reaction-bonded material maintains its strength above the melting point of the Si, further indicating that the Si has been completely converted to SiC and that the material consists of continuous SiC and not SiC grains bonded by Si, as occurs in most commercial reaction-bonded silicon carbides.

A preliminary ballistics test indicates that polymer-infiltrated SiC is much less prone to shattering than a dense SiC with similar stopping power. In these early tests, the polymer was not completely infiltrated in the reaction-bonded SiC, so it is believed that when the polymer is completely infiltrated, shattering should be reduced even further. We are building an apparatus to better catch the fragments in order to accurately quantitate shattering resistance. Infiltration with alloys is expected to give even better resistance to shattering. Figure 5 is an SEM image of completely converted porous SiC into which an aluminum alloy has been infiltrated. The gray areas are the SiC, and the white areas are the alloy. Both are continuous phases. The alloy in this case was composed of Al, Si, and Mg and was provided by Professor Martin Pech-Canul at the Unidad Saltillo of the University of Coahuila, Mexico, and the infiltration method was proposed by his group.[8] These initial tests showed good metal penetration and significant reaction between the metal and the SiC. The infiltrated SiC could not be cut with an 8-in. laboratory diamond saw because of its combined ductile and hard phases. We believe that this combination will lead to a superior armor or blast-protection material. Future work will focus on parametric testing to reduce the metal–SiC interaction and ballistic testing of the material.

Because the reaction-bonded material starts as a slurry, it can also infiltrate into woven mats of fibers that can withstand sintering temperatures. Figure 6 shows the sintered slurry interspersed among layers of carbon fiber tows. Preparation of the composites is similar to the

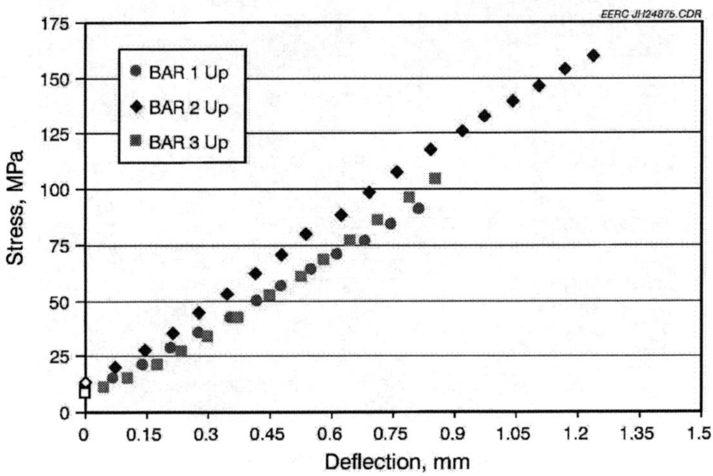

Figure 4. Four-point flexure deflection versus stress data for the reaction-bonded SiC at 1450°C.

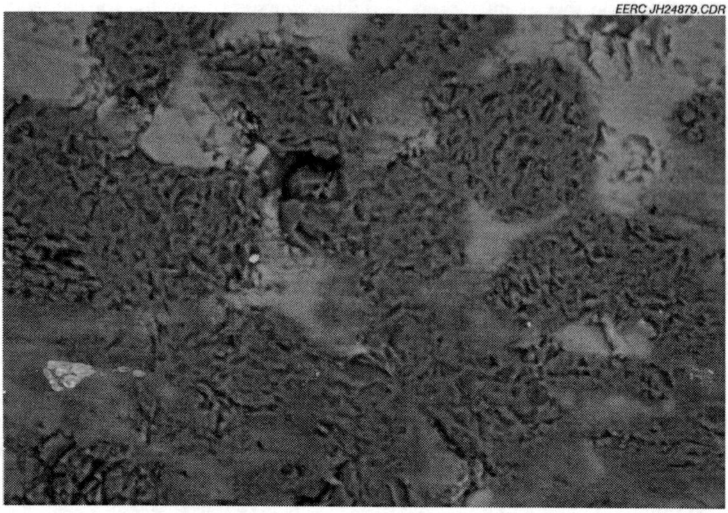

Figure 5. SEM image of continuous porous SiC into which an aluminum alloy has been infiltrated.

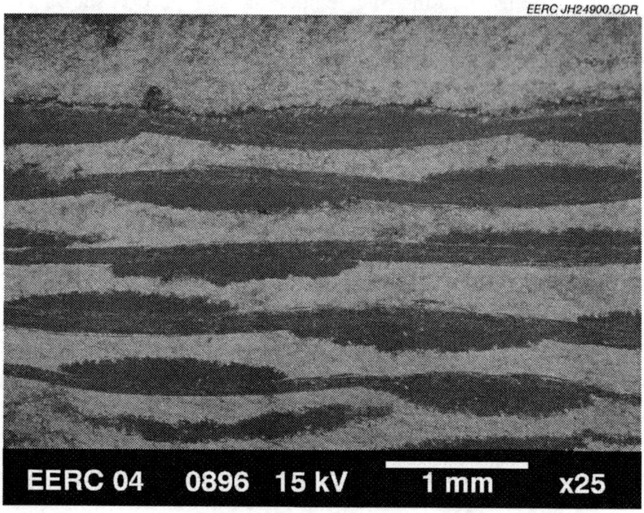

Figure 6. Sintered slurry interspersed among layers of carbon fiber tows.

fabrication of fiberglass structures. In this case, the carbon fibers outgassed during the sintering step, and the gas reacted with the surface of the Si particles, reducing their ability to sinter. We are attempting to prevent this process by outgassing the fibers before they are infiltrated with the slurry.

SUMMARY

The EERC method of creating reaction-bonded SiC involves slip-casting SiC–Si particle mixtures so that complex structures can be easily prepared, sintering the Si to the SiC, and converting the Si to continuous SiC. It does not include metal infiltration, so no residual channels of Si remain in the structure, but it is 35% porous. The porosity reduces flexural strength but permits infiltration of a wide variety of polymers or ductile metals to make significantly tougher, continuous SiC composites with high residual compressive stresses. A preliminary ballistics test indicated that polymer-infiltrated SiC is much less prone to shattering than dense SiC with similar stopping power. Metal-infiltrated SiC has not yet been tested. In addition, because the reaction-bonding material begins as a slurry, it can be infiltrated into mats of woven, high-temperature fibers to produce simple SiC–fiber composites. Production of these composites is still in the early stages.

ACKNOWLEDGMENTS

The authors would like to thank DOE NETL for funding this work under Contract Number DE-FC26-98FT40320 and Dr. Richard Read, the NETL project manager.

DISCLAIMER

REFERENCES

1. W.A. Gooch, "Ceramic Armor Materials by Design", Ceramic Transactions 134, American Ceramic Society: Westerville, OH, 3–22 (2002).

2. T.J. Moore, "Feasibility Study of the Welding of SiC." *J. Am. Ceram. Soc.* **68**, 6, C151–153 (1985).

3. T.M. Watne, J.P. Hurley, J.R. Gunderson, *Ceram. Eng. Sci. Proc.* **17**, 4A (1996).

4. T.M. Strobel, J.P. Hurley, K. Breder, J.E. Holowczak, *Ceram. Eng. Sci. Proc.* **16**, 5 (1995) 911–920.

5. K. Breder, R.J. Parten, In Proceedings of the 10th Annual Conference on Fossil Energy Materials; ORNL/CONF-9605167, pp 53–62 (1996).

6. K.V. Breder, J. Tennery, T.M. Strobel, J.P. Hurley, *J. Am. Ceram. Soc.* **78**, 10, 2837–2840 (1995).

7. B. James, "Ceramic Armor Materials by Design", Ceramic Transactions 134, American Ceramic Society: Westerville, OH, 33–44 (2002).

8. J.A. Aguilar-Martinez, M.I. Pech-Canul, M. Rodriguez-Reyes, J.L. De La Pena, *Materials Letters*, **57**, 4332–4335 (2003).

MICROSTRUCTURAL ENGINEERING OF THE Si-C-Al-O-N SYSTEM

R. Marc Flinders, Darin Ray, Angela Anderson, and Raymond A. Cutler
Ceramatec, Inc.
2425 South 900 West
Salt Lake City, Utah, 84119

ABSTRACT

Fracture mode, hardness, toughness, and sinterability are a strong function of composition for SiC-based armor compositions hot pressed with small additions of Al, AlN, Al_2O_3, or Al_4C_3 when compared to SiC-N and SiC densified with boron and carbon. Alpha SiC was used as the starting powder and the Al content was held constant while varying the amount of O, N, and C available. Rietveld analysis showed no evidence for significant solid solution formation between SiC and AlN (or Al_2OC) under the conditions used to process these materials. Oxygen was always present when processing with aluminum additives. Fracture modes varied between highly transgranular to primarily intergranular dependent on additive and processing conditions. Fracture toughness of 6.8 MPa-m$^{1/2}$ was obtained at a Vickers hardness of 20.5 GPa. Flexibility in controlling mechanical properties is discussed from a microstructural perspective.

INTRODUCTION

Silicon carbide densified with AlN[1] is the current state-of the-art armor material against heavy threats. This material produced by Cercom under the trade name SiC-N has low porosity and fractures intergranularly with toughness of 5.3 MPa-m$^{1/2}$ as measured by the single-edged precracked beam (SEPB) technique.[2] SiC densified with B and C fractures transgranularly with higher hardness and half the SEPB toughness.[3] While intergranular fracture is one of the keys for obtaining high toughness in SiC, testing against 14.5 mm WC-Co showed little difference in V_{50} performance for 12.7 mm thick tiles having different fracture modes.[4] Lundberg and Lundberg[5] recently reported that the transition velocity from interface defeat to penetration for long rods correlated better with increases in indentation toughness than Vickers hardness. Because ballistic performance is threat dependent it is important to determine how different microstructures influence dwell and penetration. Pickup[6] has suggested based on Kolsky bar testing that dwell changes more than penetration as microstructure is varied from intergranular to transgranular fracture. It is possible to control the fracture mode by the choice of additive and processing conditions for a variety of SiC ceramics.

Alliegro, et al.[7] found that small additions of Al and/or Fe allowed SiC to be hot pressed to near theoretical density. Prochazka[8] pioneered the commercial use of SiC by demonstrating pressureless sintering using B and C additions. Small additions of Al and C or Al, B, and C were used for pressureless sintering[9,10]. When Al, B, and C are used to densify SiC under applied pressure, higher additions of Al increase the fracture toughness by changing the fracture mode from transgranular to intergranular fracture.[11] Intergranular fracture and toughening is also observed when Al_2O_3 is used as an additive.[12,13] Cutler, et al.[14] used the fact that AlN and Al_2OC are isostructural with 2H SiC in order to make SiCAlON solid solutions. Additions of AlN or Al_2OC to SiC reduced the grain size and promoted intergranular fracture.[15,16] Al_4C_3 and Al_2O_3 were used to make in-situ Al_2OC, which promoted sintering by liquid phase sintering.[17] Liquid phase sintering with Al_2O_3 and Y_2O_3 also results in intergranular fracture and modest improvements in toughness.[18] Elongation of grains using AlN and Y_2O_3 as sintering aids further increases the fracture toughness of this material.[19,20] AlN and yttrium aluminum garnet (YAG) are additives to commercially available SiC densified using a liquid phase.[21]

It was desired to compare solid-state sintered SiC densified with B and C to SiC hot pressed using AlN, Al, or Al_2O_3 at the same aluminum level. It was speculated that the oxygen content of the starting Al-based powders would produce some Al_2O_3 at grain boundaries and/or triple points so that intergranular fracture could occur as compared with SiC densified with B and C, where the C removes the surface silica. Al_4C_3 was chosen as an additional additive since it was postulated that the close association of Al and C would permit the removal of O from the additive prior to decomposition and might lead to transgranular fracture at the same level of aluminum. The purpose of this paper is to compare Al-based additives which allow excellent densification of SiC under applied pressure and explore how changes in microstructure affect room-temperature hardness and fracture toughness.

EXPERIMENTAL PROCEDURES

The starting powders used for processing are listed in Table I along with data provided by the suppliers. All compositions were prepared by batching 600 grams of powder in two-liter high-density polyethylene (HDPE) jars filled with 1.6 kg solid state sintered SiC media and 700 g reagent grade acetone. The slurries were mixed for 16 hours on a ball mill in order to disperse the agglomerates and the powders were stir dried before screening through an 80-mesh screen. The SiC batched with B_4C and C was pyrolyzed by heating in N_2 to 600°C. Billets (45 mm x 45 mm x 6 mm) were hot pressed at 28 MPa inside graphite dies by heating to 1500°C in a vacuum of 1 torr, holding at 1500°C for one hour to remove SiO and CO from the samples, and then backfilling with Ar or N_2 and heating to temperatures between 1900 and 2200°C for one hour.

Cercom's hot pressed SiC-N was supplied by the Army Research Laboratory (ARL). The hot pressed billets were ground with a 320 grit diamond wheel to make 3 mm x 4 mm x 45 mm bars as specified by ASTM C-1421-99.[22] Toughness was measured using the single-edge precracked beam (SEPB) technique using black printer ink to mark cracks as described previously.[2] All crack planes were parallel to the hot pressing direction. Each data point is the mean of five bars tested, with error bars representing one standard deviation.

A microhardness machine (Leco model LM-100) was used to obtain Vickers and Knoop hardness data on polished SEPB bars. Data were taken at a load of 9.8 N. Each data point represents the mean of ten measurements, with error bars representing one standard deviation.

Table I
Raw Materials Used in Powder Processing

Powder	Supplier	Grade	Information from Supplier*
α-SiC	H. C. Starck	UF-15	O=1.0%, C=29.6%, d_{50}=0.5 μm, SA=15 m^2/g
Al	Valimet	H-3	Fe=0.12%, Al=99.8%, d_{50}=5.0 μm
AlN	Tokuyama Soda	F	O=0.78%, C=0.03%, SA=3.4 m^2/g
Al_2O_3	Sasol North Am.	SPA-0.5	Si=14 ppm, d_{50}=0.47 μm, SA=7.6 m^2/g
Al_4C_3	H. C. Starck	---	O=1.7%, d_{50}=3.85 μm
B_4C	H. C. Starck	HS	O=1.5%, B/C=3.80, d_{50}=0.96 μm, SA=18.0 m^2/g
C	Capital Resin	CRC-720	40 wt. % solids after pyrolysis

*wt. % unless otherwise noted

Rietveld analysis[23,24] was used to determine SiC polytypes present in the densified samples with X-ray diffraction patterns collected from 20-80° 2-theta, with a step size of 0.02°/step and a counting time of 4 sec/step.

Polished surfaces were etched with boiling Murakami's reagent (10 g KOH, 10 g $K_3Fe(CN)_6$ and 30 g H_2O) for times ranging from 2 to 15 minutes dependent on composition and grain size. Grain size was determined by the line-intercept method, where the multiplication constant ranged between 1.5 (equiaxed grains) and 2.0 (elongated, plate-shaped grains).[25] Typically, 200-300 grains were measured for each composition in order to get a mean grain size.

The aspect ratios of the five most acicular grains in each of three micrographs were used to estimate a comparative aspect ratio. The fracture mode was determined from broken SEPB bars. Oxygen and nitrogen were analyzed on selected starting powders and sintered samples using ASTM method E1409 for the determination of oxygen by the inert gas fusion/thermal conductivity detection technique with a Leco TC-136 model analyzer with detection limits below 5 wt. %.

RESULTS AND DISCUSSION

Choice of Sintering Additive

Figure 1 shows XRD scans for four compositions of α-SiC with equimolar additions of aluminum, added as Al metal (1.65 wt. %), Al_4C_3 (2.2 wt. %), AlN (2.50 wt. %), or Al_2O_3 (3.10 wt. %), and hot pressed at 2100°C for 1 hour. Included for comparison are a solid state SiC (SiC densified with 2.0 wt. % C and 0.5 wt. % B_4C), hot pressed under the same conditions, and commercially available SiC-N. Since all materials started with primarily 6H SiC, it is possible to see how the different additives influence the polytypes in the densified materials. As shown in Table II, AlN additions did not produce any 2H polytypes. Based on the work of Zangvil and Ruh[26] one could expect a 4H solid solution to occur between SiC and AlN. Since 6H is still the primary phase present it is difficult to conclude, based on XRD, that any solid solution has occurred. Aluminum is known[27] to promote the 6H to 4H transformation, consistent with the results shown in Table II for the SiC-Al sample. Al_2O_3 and Al_4C_3 produced similar results in polytypes, contrary to expectation since it was expected that aluminum carbide would decompose to give Al and C. If the C removed the surface oxygen for the Al_4C_3 then one would expect that Al_4C_3 or Al_4SiC_4, if formed, would decompose by 2100°C to give free Al and C.[28] No Al_4SiC_4, Al, C, or Al_4C_3 secondary phases were detected by XRD for the sample made with aluminum carbide, though some free carbon was observed in the microstructure by SEM. The samples with the most 6H SiC were the solid state and the SiC-N samples. Since the starting powder for the SiC-N is likely different from the other samples, it is only included for comparison.

Oxygen and nitrogen analysis in the densified materials is also shown in Table II. The starting lot of SiC was analyzed to have 1.2 wt. % oxygen. The Al starting powder was measured at 0.76 wt. % O and the AlN at 1.3 wt. % O. The Al_4C_3, surprisingly, had an oxygen level above the detection limit of 5 % as well as 0.9 wt. % N. The Al_2O_3 powder contained 0.11 wt. % N, in addition to its theoretical 31.4 wt. % O. As shown in Table II, the phenolic resin in the solid-state material was effective in removing the oxygen, but whenever an Al compound was present there was significant oxygen present. Al was more effective than Al_2O_3 in retaining oxygen, and the stability of the Al_4C_3 apparently prevented it from allowing its carbon to remove the O associated with it during the 1500°C hold when the samples had open porosity.

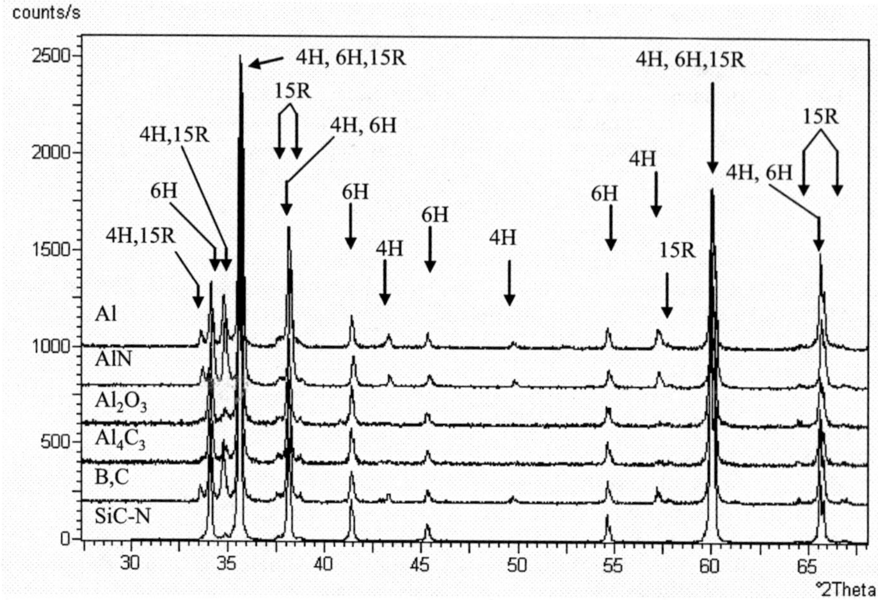

Figure 1. X-ray diffraction scans of SiC hot pressed at 2100°C with Al-based additives compared with solid state SiC hot pressed at the same temperature and Cercom's SiC-N.

Minimal nitrogen was retained by the Al_4C_3 suggesting that adsorbed species were likely present in this material. In contrast, samples made using AlN retained N in the sample, suggesting that some solid solution formation had occurred by this temperature.

The etched microstructures in Figure 2 show the case-core structure in both the SiC-AlN and the SiC-N samples, observable when the nitrogen, with a lower diffusion coefficient, only partially enters the structure. Schwetz, et al.[21] showed that the case contains both N and O in solid solution with the SiC. In addition, energy dispersive spectroscopy (EDS) in conjunction with TEM samples has shown both aluminum and oxygen at the triple junctions in compositions containing Al.[29]

Table II
Density, Chemical Analysis and Rietveld Phase Assemblage

Composition	H.P. Temp. (°C)	Density (g/cm^3)	Chemical Analysis (wt%) O	N	SiC Polytypes (wt. %) 4H	6H	15R
1.65 wt. % Al	2100	3.21	0.76	0.008	61.7	33.0	5.4
2.50 wt. % AlN	2100	3.22	0.44	0.20	26.8	62.2	7.7
3.10 wt. % Al_2O_3	2100	3.21	1.08	0.004	9.7	83.1	7.2
2.2 wt. % Al_4C_3	2100	3.21	0.67	0.015	7.8	83.5	8.7
0.5 wt. % B_4C-2 wt. % C	2100	3.19	0.02	0.008	0.0	94.8	5.2
Cercom SiC-N	-	3.22	0.56	0.17	0.0	100	0.0

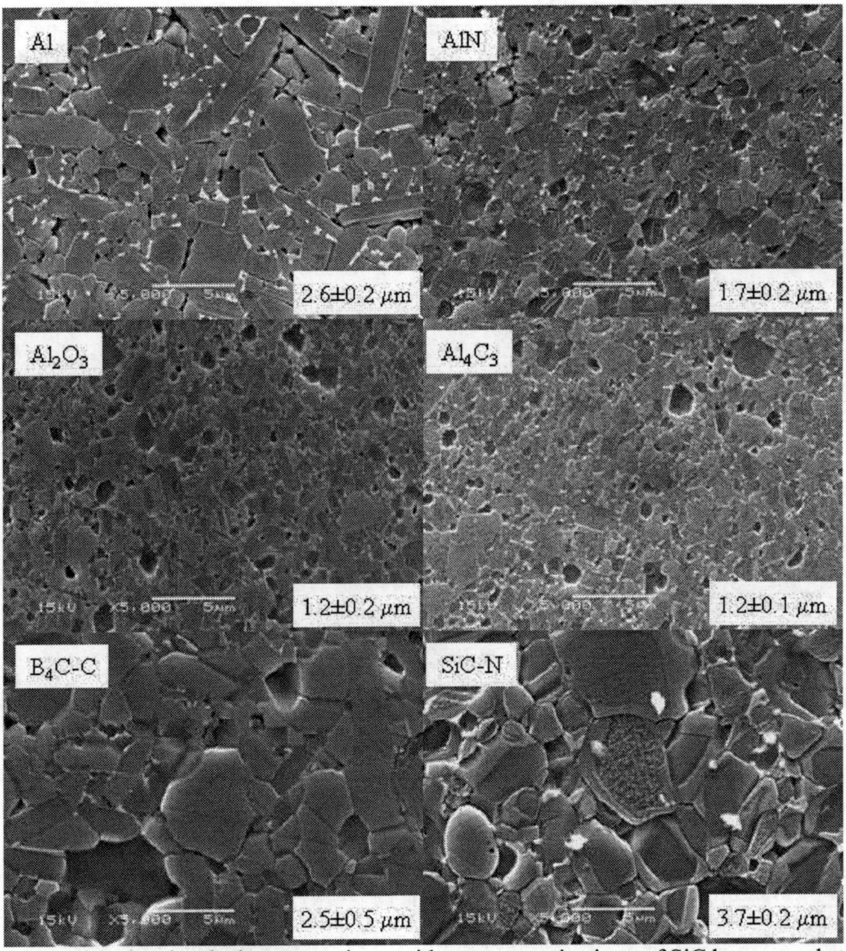

Figure 2. Polished and etched cross-sections, with average grain sizes, of SiC hot pressed at 2100°C with various Al-based additives compared with solid-state SiC and SiC-N (see Table II).

The presence of nitrogen slows down grain growth consistent with the early work of Rafaniello.[30] The SiC-N material is much coarser than the SiC-2.5 wt. % AlN sample, even though the nitrogen levels in the sintered materials were similar. The grain size of the SiC-N is comparable to the solid state sintered sample. The samples prepared with Al had comparable density to those prepared with AlN, but showed elongated grains even though the starting powder was α-SiC powder. While the Al_2O_3 and Al_4C_3 show some plate growth, the microstructures are generally equiaxed and fine grained.

Fracture surfaces are displayed along with SEPB toughness values in Figure 3. The only sample with primarily transgranular fracture was the solid-state material, which accounts for its low toughness.[3] The Al material had fracture toughness similar to SiC-N even though the microstructures are very different. In the case of Al, the fracture mode is mixed, whereas the SiC-N is primarily intergranular. The materials processed with AlN, Al_2O_3, or Al_4C_3 have intermediate toughness values, due to their intergranular fracture mode but much finer grain size.[2,3]

An alternate method of observing fracture mode is to examine the path of the cracks generated from Vickers indents made on polished cross-sections, as is shown in Figure 4. Intergranular or mixed mode fracture of coarse-grained materials shows cracking around indents and crack deflection and grain bridging as the cracks extend away from the corners of the indents.[11]

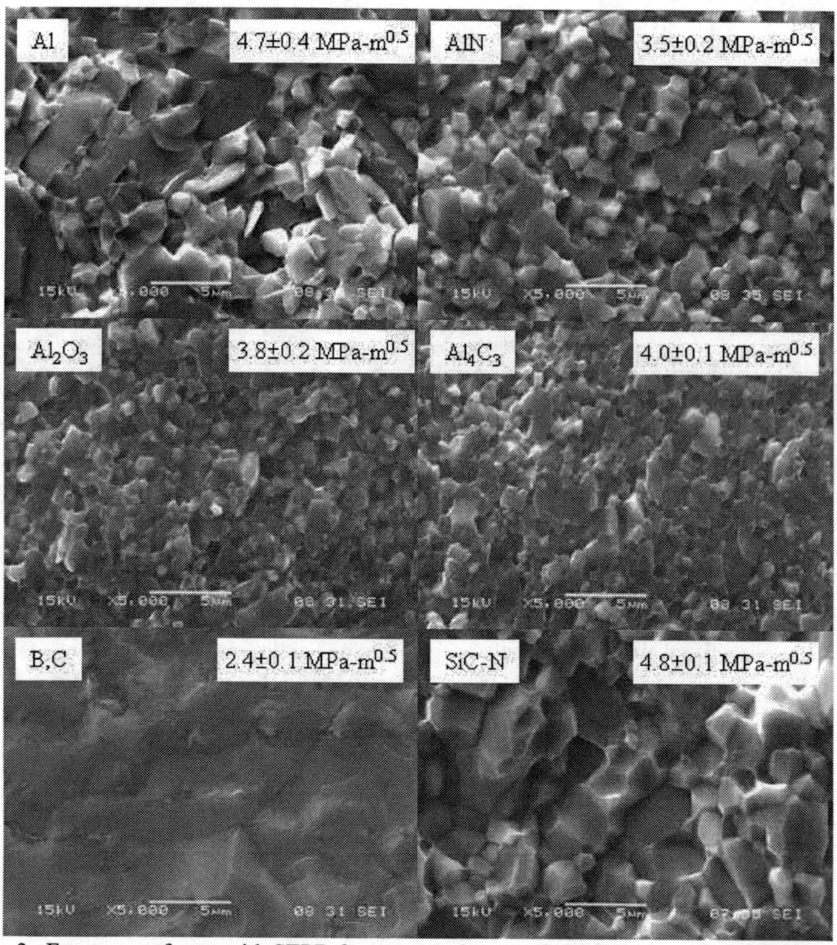

Figure 3. Fracture surfaces, with SEPB fracture toughness, of SiC hot pressed at 2100°C with various Al-based additives compared with solid-state SiC and SiC-N (see Table II).

This is most apparent for SiC-N and SiC-1.65 wt. % Al shown in Figure 4, though all of the SiC made with aluminum-based additives show this to a certain extent, which is readily apparent when these indents are compared with the solid-state material. While hardness numbers in the literature are often difficult to compare due to differences interpreting indent diagonals, it is clear that the hardness of SiC densified with Al is superior to that densified with Al, B, and C.[11] Though this is partially due to better densification, it also appears that the grain boundary phase affects hardness.[3] All of the materials appear to have hardness as good as or better than SiC-N. When both hardness and toughness are considered, it is apparent that a range of properties can be produced by changing the sintering additive.

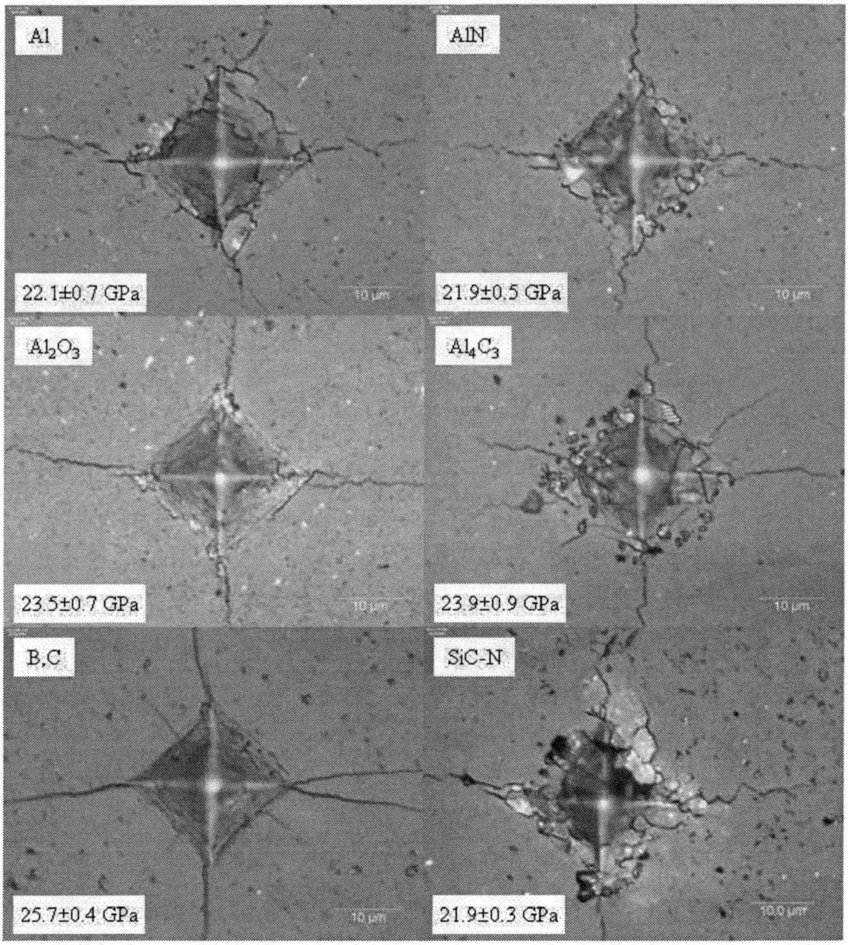

Figure 4. Polished cross-sections, with Vickers hardness at 1 kg load, of SiC hot pressed at 2100°C with various Al-based additives compared with solid-state SiC and SiC-N (see Table II).

Processing Temperatures

The densification temperature affects properties primarily due to differences in grain size. Table III lists properties of the same compositions processed at different temperatures. Figure 5 shows the hardness-toughness trade-off that occurs in the SiC-1.65 wt. % Al system as a function of hot pressing temperature. While fracture toughness has been shown to be a strong function of grain size, the observed hardness dependence on grain size is in stark contrast to the ABC-, YAG-, and solid-state SiC systems which has shown little or no grain size dependence.[3] The difference may be the fracture mode, which remains mixed intergranular/transgranular in the processing temperature range, as shown in Figure 6. Further work in quantifying the fracture mode would be needed to determine if an increase in transgranular fracture character was responsible for the increase in hardness. Alternatively, hardness may in fact be a function of grain size for SiC fracturing intergranularly, but only when the grain size is sufficiently smaller than the Vickers indent size, as suggested by Rice et al.[31] Previous work[2,3] by the authors done in other SiC systems dealt with SiC materials in larger grain size ranges.

As can be seen by the data in Table III, an increase in hot pressing temperature leads to a reduction in both oxygen and nitrogen content. The problem with processing at high temperatures is exaggerated grain growth, which is common with solid-state sintering. This was also observed when Al_2O_3 and Al_4C_3 were used as additives, but not when Al or AlN were the sintering additive. The amount of oxygen generally decreases with increasing temperature, which suggests that closed porosity was not obtained until the hold temperature was reached. The Al additive was the only one that allowed densification at 1900°C to a density of greater than 3.20 g/cc. Al appears to be an attractive choice when using pressure-assisted densification.

Amount of Additive

Table IV gives characterization data for a range of additives at different concentrations and sintering temperatures. When considering the amount of additive in the Al- or AlN-

Table III

Effect of Hot Pressing Temperature on Properties

Comp.	H.P. Temp. (°C)	Density (g/cm³)	Hardness Vickers (GPa)	Hardness Knoop (GPa)	SEPB Toughness (MPa-√m)	Chemical Analysis (wt%) O	Chemical Analysis (wt%) N	SiC Polytypes (wt. %) 4H	6H	15R
1.65 Al	1900	3.22	25.3±0.7	20.1±0.2	4.0±0.2	1.01	0.006	8.1	83.5	8.5
	2000	3.21	24.2±0.3	19.7±0.1	4.1±0.1	1.16	0.010	31.7	59.9	8.5
	2100	3.21	22.1±0.7	19.2±0.2	4.7±0.4	0.76	0.008	61.7	33.0	5.4
	2200	3.21	20.8±0.3	18.9±0.2	5.7±0.1	0.46	0.006			
2.50 AlN	2100	3.22	21.9±0.5	19.8±0.3	3.5±0.2	0.44	0.20	26.8	62.2	7.7
	2200	3.21	22.9±0.1	19.7±0.2	3.5±0.4	0.64	0.42	39.7	53.0	7.2
3.10 Al_2O_3	2000	3.17	22.0±1.1	20.0±0.1	3.6±0.1	1.50	0.009	0.0	89.4	10.6
	2100	3.21	23.5±0.7	20.7±0.1	3.8±0.2	1.08	0.004	9.7	83.1	7.2
	2200	3.21	23.2±0.3	19.3±0.3	4.1±0.2	0.49	0.008			
2.2 Al_4C_3	2000	3.22	22.7±0.5	20.2±0.2	3.9±0.1	1.22	0.023	7.6	83.7	8.7
	2100	3.21	23.9±0.9	20.5±0.2	4.0±0.1	0.67	0.015	7.8	83.5	8.7
	2200	3.21	21.7±0.4	20.0±0.3	3.6±0.3	0.65	0.021	13.6	78.4	8.0
0.5B_4C-2C	2100	3.19	25.7±0.4	20.8±0.1	2.4±0.0	0.02	0.008	0.0	94.8	5.2
	2200	3.20	25.6±1.0	21.2±0.2	3.3±0.3			28.6	71.4	0.0
SiC-N	-	3.22	22.8±0.4	20.3±0.2	4.8±0.1	0.56	0.17	0.0	100	0.0

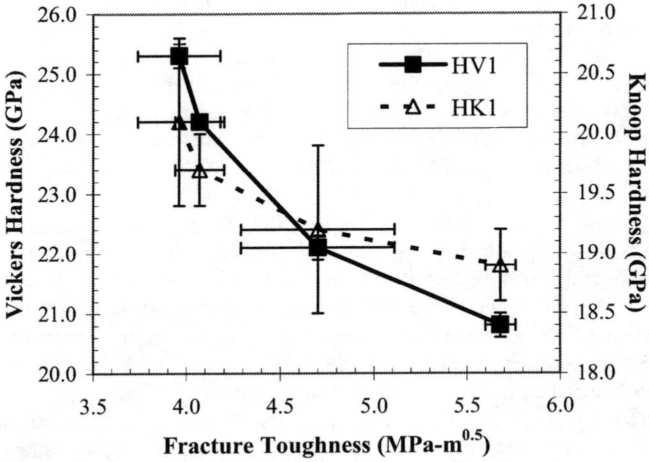

Figure 5. Hardness-toughness trade-off for SiC-1.65 wt. % Al densified at temperatures ranging between 1900 and 2200°C. Toughness increases with grain size, while hardness decreases.

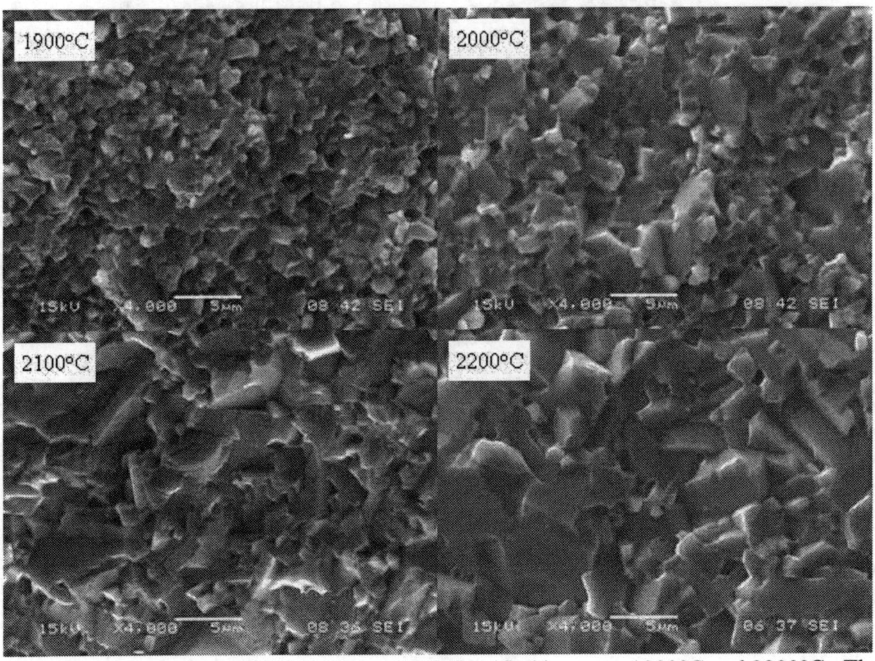

Figure 6. Fracture surfaces of SiC-1.65 wt. % Al densified between 1900°C and 2200°C. The fracture mode remains mixed at all temperatures.

compositions, the additive content was controlled such that it was doubled or cut in half relative to compositions discussed above.

The amount of additive also affects the mechanical properties. In the case of AlN, increasing the amount of AlN decreases the grain size thereby increasing the hardness as shown in Figure 7. This is consistent with the work of Rafaniello[30] where he found grain refinement in combination with 2H solid solution formation. Since the amount of nitrogen increases with increasing nitrogen content and is retained to 2200°C, it is likely that some solid solution formation is occurring.

The AlN system is very interesting in that it both promotes densification at lower temperatures than with solid-state sintering and retards grain growth. It is likely the interplay between O at grain boundaries with dissolved O and N with SiC grains that allows this to occur. The fracture mode in the SiC-5 wt. % AlN changes with temperature as shown in Figure 8. At 2200°C it is mixed changing to transgranular at 2000°C. This change may show the influence of solid solution formation on fracture mode. At 1900°C the facture mode is intergranular, but this is most certainly influenced by the low density of this material (see Table IV).

The fracture toughness was relatively invariant with grain size and amount of additive for the SiC-AlN compositions, with the microstructure remaining essentially equiaxed, and all values of toughness near 4 MPa-m$^{1/2}$. The hardness, however, was influenced by grain size with higher AlN giving hardness similar to solid-state SiC. In contrast the SiC-Al system, as shown in Figure 9, allowed a more apparent grain growth, with an elongated microstructure and a higher toughness (6.8 MPa-m$^{1/2}$) with more additive, but with the expected trend in reduced hardness.

Table IV
Effect of Amount of Additive on Properties

Comp.	H.P. Temp. (°C)	Density (g/cm^3)	Hardness Vickers (GPa)	Hardness Knoop (GPa)	SEPB Toughness (MPa-√m)	Chemical Analysis (wt%) O	Chemical Analysis (wt%) N	SiC Polytypes (wt. %) 4H	SiC Polytypes (wt. %) 6H	SiC Polytypes (wt. %) 15R
0.83 Al	1900	3.01								
(0.5x)	2000	3.22	23.4±0.8	19.9±0.2	4.0±0.1			2.3	88.6	9.0
	2100	3.22	23.5±0.9	20.1±0.2	3.8±0.1			6.4	86.2	7.4
3.30 Al	1900	3.18	23.4±0.9	25.5±0.1						
(2x)	2000	3.19	22.1±0.6	18.9±0.2	4.2±0.1	0.80	0.010	12.7	79.7	5.9
	2100	3.19	19.9±0.4	17.8±0.1	4.7±0.6	0.37	0.007	28.1	66.0	4.6
	2200	3.18	20.5±0.5	18.5±0.1	6.8±0.1	0.27	0.007			
1.25 AlN	2100	3.22	22.7±0.1	20.3±0.1	4.0±0.1			7.2	84.8	8.0
(0.5x)	2200	3.21	22.4±0.8	20.1±0.1	4.1±0.1	0.45	0.16	0.0	91.8	8.2
5.00 AlN	1900	2.46								
(2x)	2000	3.19	25.0±0.4	21.2±0.2	3.9±0.2	0.72	1.43	0.8	89.5	9.7
	2100	3.23	24.2±0.5	21.3±0.1	3.8±0.1	0.75	1.22	0.0	90.2	9.8
	2200	3.22	24.1±0.6	20.9±0.2	4.2±0.1	0.49	0.89	6.6	82.5	10.9
0.83 Al – 1.25 AlN	2100	3.22	23.2±0.4	20.4±0.2	4.1±0.1	0.79	0.36	17.0	74.0	8.9
1.65 Al (run in N_2)	2100	3.21	22.4±0.3	20.9±0.2	4.3±0.1	0.77	0.36	41.7	53.0	5.4
3.30 Al (run in N_2)	2100	3.18	19.4-24.7	18.4-20.5	4.3±0.8	1.13	0.45	27.6	65.9	5.4

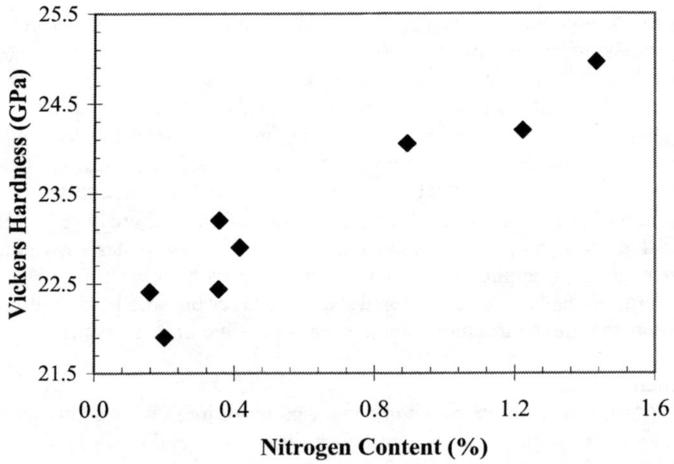

Figure 7. Effect of nitrogen content in SiC-AlN samples on hardness. Note that nitrogen increases hardness by decreasing grain size.

Figure 8. Fracture surfaces of SiC-5 wt. % AlN densified at temperature ranging between 1900 and 2200°C. Note how fracture mode changes with temperature.

Influence of Atmosphere

Nitrogen can be added to SiC as a solid (such as with AlN or Si_3N_4), or can be introduced as a gas during processing. To demonstrate the latter, SiC-1.65 wt. % Al and SiC-3.30 wt. % Al compositions were backfilled with N_2 rather than Ar at the end of the 1500°C vacuum hold. These materials contained nitrogen levels comparable to their AlN counterparts, as shown in Table IV. The nitrogen was not uniformly distributed within the material as shown in Figure 10. This clearly shows that the edges of the sample were fine-grained and hard, similar to the SiC-AlN material, and the core was elongated and tougher, like SiC-Al, due to the gradient in nitrogen concentration. An extended hold, or the absence of applied pressure at 1500°C, while the body was still porous, may have allowed the nitrogen to be better distributed. It may however be advantageous in applications such as ceramic armor, to have a functionally graded material, with a harder outside layer co-sintered with a tougher but less hard core material. It is clear by these results that microstructural engineering is possible in these systems.

Influence of Carbon and Boron

Table V lists a select number of companion compositions, with additions of boron (as boron carbide) and/or free carbon. It was anticipated that boron additions would promote grain growth and could lead to higher oxygen contents while carbon and boron additions would results

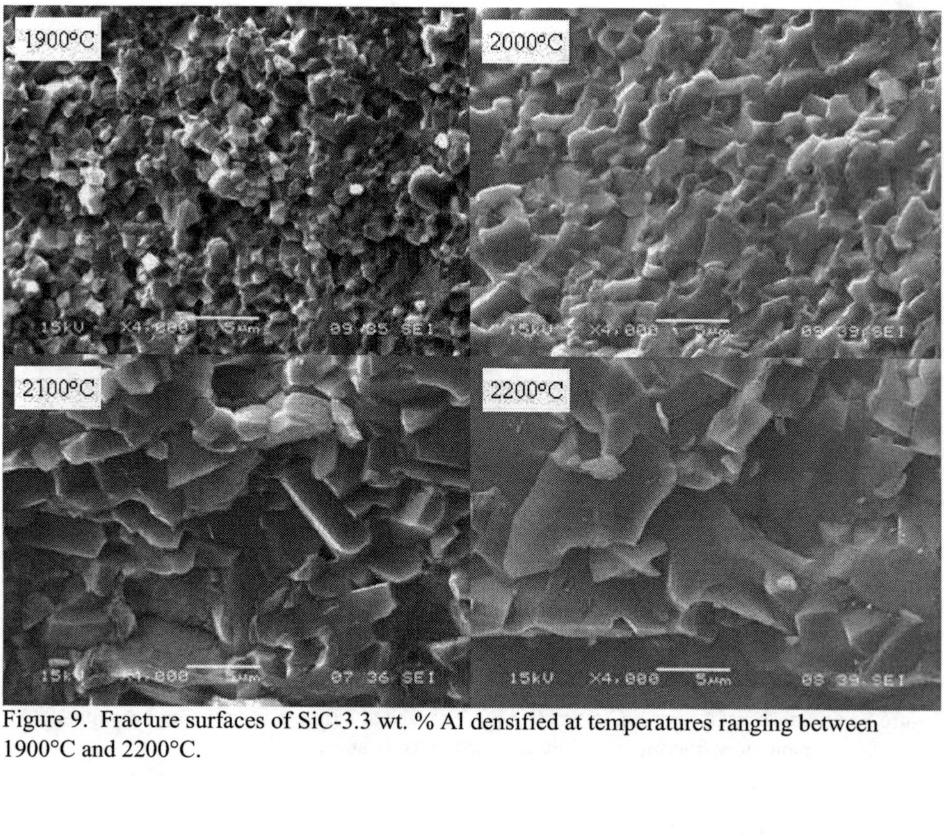

Figure 9. Fracture surfaces of SiC-3.3 wt. % Al densified at temperatures ranging between 1900°C and 2200°C.

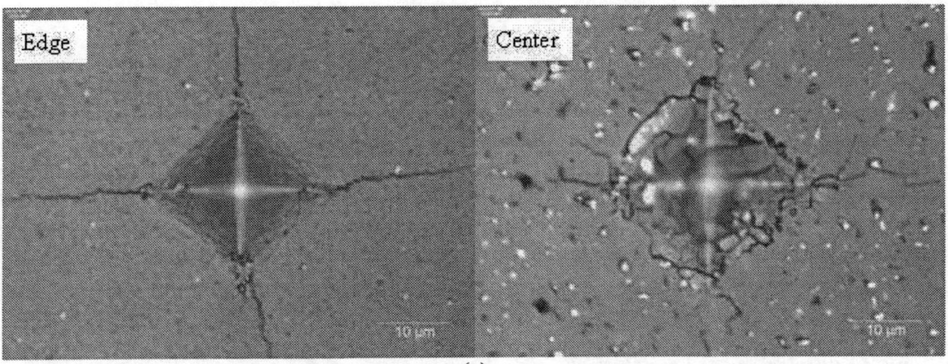

(a)

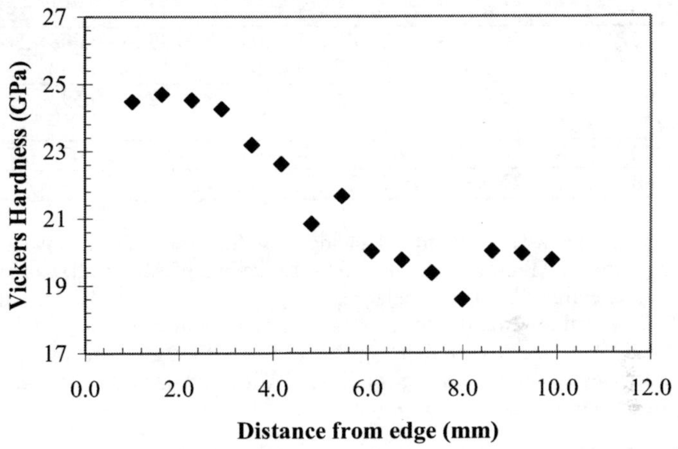

(b)

Figure 10. HV1 hardness of SiC-3.3 wt. % Al hot pressed in N_2. (a) Vickers indents from edge and core showing change in fracture mode, and (b) increase in hardness as nitrogen enters lattice.

in grain growth with lower oxygen content due to the phenolic resin helping to remove surface oxygen from the Al-based additives, as well as the surface silica on the SiC. By comparing the data in Tables III, IV, and V it is possible to see that more oxygen and nitrogen remained in Al-containing materials sintered with B_4C or B_4C+C. Boron and carbon likely speed up the densification kinetics allowing the materials to reach a state of closed porosity earlier. For the solid state material, however, the oxygen level was very low since there was no Al to getter oxygen and the kinetics of densification were slow enough to permit removal of the O as SiO and CO.

The consequence of grain growth is apparent when comparing the microstructures of SiC-2.5 wt. % AlN and SiC-2.5 wt. % AlN-2 wt. % C-0.5 wt. % B_4C hot pressed at 2200°C as

Table V
SiC Compositions with Additional Additives

Comp.	H.P. Temp. (°C)	Density (g/cm^3)	Hardness Vickers (GPa)	Hardness Knoop (GPa)	SEPB Toughness (MPa-√m)	Chemical Analysis (wt%) O	Chemical Analysis (wt%) N	SiC Polytypes (wt. %) 4H	SiC Polytypes (wt. %) 6H	SiC Polytypes (wt. %) 15R
w/ 0.5 wt. % B_4C										
1.65 Al	2200	3.16	22.4±0.1	20.0±0.3	3.7±0.1	0.25	0.008			
2.50 AlN	2100	3.20	21.0±0.6	18.7±02	3.4±0.4			52.3	44.0	3.7
	2200	3.20	23.1±0.4	20.3±0.2	2.7±0.2	0.38	0.84			
5.00 AlN	2100	3.20	22.7±0.5	19.9±0.3	3.7±0.2					
	2200	3.21	21.4±0.6	19.5±0.2	5.8±01	0.56	1.27	57.8	30.2	12.0
3.10 Al_2O_3	2200	3.19	23.2±0.7	20.8±0.2	3.0±0.3					
w/ 0.5 wt. % B_4C – 2.0 wt. % C										
No Al	2100	3.19	25.7±0.4	20.8±0.1	2.4±0.0	0.02	0.008	0.0	94.8	5.2
	2200	3.20	25.6±1.0	21.2±0.2	3.3±0.3			28.6	71.4	0.0
1.65 Al	2100	3.16	20.3±0.3	18.2±0.5	3.1±0.1	0.98	0.008	21.2	71.7	7.1
	2200	3.16	22.2±0.4	19.0±0.2	3.4±0.1	0.73	0.008	62.7	31.6	5.7
2.50 AlN	2100	3.16	22.7±0.3	19.0±0.2	2.8±0.2	0.72	0.59	49.4	45.7	4.9
	2200	3.16	20.6±0.8	18.7±0.1	3.6±0.1					
5.00 AlN	2200	3.18	20.3±0.5	19.0±0.3	4.8±0.4	0.82	1.67	27.7	64.1	8.3
3.10 Al_2O_3	2200	3.17	22.0±0.8	20.1±0.3	2.9±0.1					

shown in Figure 11. The faster grain growth changes the fracture mode and lowers the toughness from 3.5 to 2.7 MPa-$m^{1/2}$. By doubling the amount of AlN the fracture mode can be changed to mixed intergranular/transgranular resulting in a toughness of 5.8 MPa-$m^{1/2}$. Carbon additions of 2 wt. % lead to lower densities, thereby decreasing modulus and hardness (compare Tables III, IV, and V). It is apparent that there are a wide variety of microstructures possible by controlling additive levels and processing conditions. Ultimately, ballistic data are needed to assess what type of microstructure leads to improved penetration resistance.

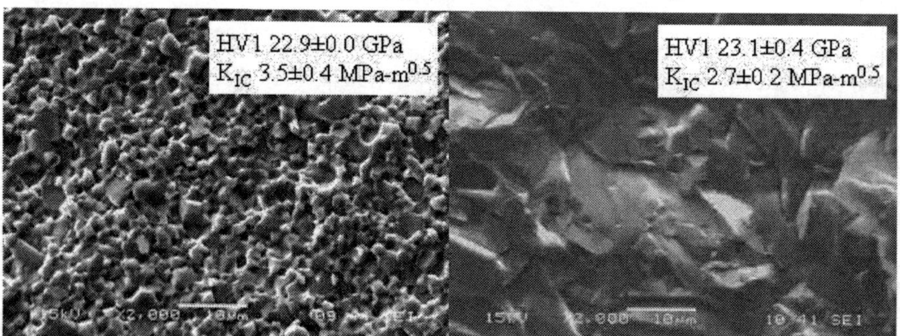

Figure 11. Comparison of fracture surfaces for SiC-2.5 wt. % AlN densified at 2200°C for one hour. Note change in fracture mode for intergranular (left) to transgranular (right) by the addition of 0.5 wt. % B_4C.

CONCLUSIONS

SiC was hot pressed to densities exceeding 3.20 g/cc using Al, AlN, Al_2O_3 or Al_4C_3 as additives. Hardness comparable to solid-state densified SiC, with improvement in toughness, was obtained by keeping the grain size small in SiC-1.65 wt. % Al or 5.0 wt. % AlN. The toughness of solid state SiC could be at least doubled with other Al or AlN as additives, with a 15 to 20 % decrease in HV1 hardness. The most important finding of this work is the realization that a rich interplay of microstructure and mechanical properties is possible by controlling chemistry. Nitrogen, for example, slows down grain growth increasing hardness and decreasing toughness. It is anticipated that changes in microstructure will affect ballistic performance at a given threat level.

ACKNOWLEDGEMENT

This SBIR work was performed for the U.S. Army under contract DAAD17-02-C-0052. Appreciation is expressed to Lyle Miller of Ceramatec for help with x-ray diffraction work. Technical discussions with Dr. Svante Prochazka (retired GE), Dr. William Rafaniello, and Dr. Jim Campbell (ARL) are gratefully acknowledged.

REFERENCES

[1]A. Ezis, "Monolithic, Fully Dense Silicon Carbide Material, Method of Manufacturing, and End Uses," U. S. Patent 5,372,978 (Dec. 13, 1994).

[2]D. Ray, M. Flinders, A. Anderson, and R. A. Cutler, "Hardness/Toughness Relationship for SiC Armor," *Ceram. Sci. and Eng. Proc.*, **24** 401-10 (2003).

[3]M. Flinders, D. Ray and R. A. Cutler, "Toughness-Hardness Trade-Off in Advanced SiC Armor," *Ceram. Trans.*, **151** 37-48 (2003).

[4]D. Ray, M. Flinders, A. Anderson, R. A. Cutler, and W. Rafaniello, "Effect of Room-Temperature Hardness and Toughness on the Ballistic Performance of SiC-Based Ceramics," to appear in *Ceram. Sci. and Eng. Proc.*, **26** (2005).

[5]P. Lundberg and B. Lundberg, "The Transition Between Interface Defeat and Penetration for Tungsten Projectiles and Four Silicon Carbide Materials," *Int. J. Impact Eng.*, **31** 781-795 (2005).

[6]I. Pickup, "The Correlation of Microstructure and Mechanical Characteristics of Silicon Carbide with Ballistic Performance," to appear in *Ceram. Sci. and Eng. Proc.*, **26** (2005).

[7]R. A. Alliegro, L. B. Coffin, and J. R. Tinklepaugh, "Pressure-Sintered Silicon Carbide," *J. Am. Ceram. Soc.*, **39**[11] 386-89 (1956).

[8]S. Prochazka, "Role of Boron and Carbon in the Sintering of Silicon Carbide," *Proc. of the Conference on Ceramics for High Performance Applications*, Hyannis, Mass., 1973, ed. by J. J. Burke, A. E. Gorum, and R. M. Katz (Brook Hill Publ. Co., 1975).

[9]W. Böcker, H. Landfermann, and H. Hausner, "Sintering of Alpha Silicon Carbide with Additions of Aluminum," *Pow. Met.*, **11** 83-85 (1979).

[10]K. A. Schwetz and A. Lipp, "The Effect of Boron and Aluminum Sintering Additives on the Properties of Dense Sintered Alpha Silicon Carbide," *Science of Ceramics,* **10** 149-158 (1980).

[11]M. Flinders, D. Ray, A. Anderson, and R. A. Cutler, "High-Toughness Silicon Carbide as Armor," to appear in *J. Am. Ceram. Soc.*, **88** (2005).

[12]F. F. Lange, "Hot-Pressing Behavior of Silicon Carbide Powders with Additions of Aluminum Oxide," *J. Mater. Sci.*, **10** 314-20 (1975).

[13]K. Suzuki, "Pressureless Sintering of Silicon Carbide with Addition of Aluminum Oxide," Asahi Garasu Kenkyu Hokoku, **36**[1] 25-36 (1986).

[14]I. B. Cutler, P. D. Miller, W. Rafaniello, H. K. Park, D. P. Thompson and K. H. Jack, "New Materials in the Si-C-Al-O-N and Related Systems," *Nature*, **275** 434-35 (1978).

[15]W. Rafaniello, K. Cho and A. Virkar, "Fabrication and Characterization of SiC-AlN Alloys," *J. Mater. Sci.*, **16** 3479-88 (1981).

[16]T. B. Jackson, A. C. Hurford, S. L. Bruner and R. A. Cutler, "SiC-Based Ceramics with Improved Strength," pp. 227-40 in Silicon Carbide '87, Edited by J. D. Cawley and C. E. Semler (Am. Ceram. Soc., Columbus, OH.,1989).

[17]J. L. Huang, A. C. Hurford, R. A. Cutler and A. V. Virkar, "Sintering Behavior and Properties of SiCAlON Ceramics," *J. Mater. Sci.*, **21** 1448-56 (1986).

[18]R. A. Cutler and T. B. Jackson, 'Liquid Phase Sintered Silicon Carbide," pp. 309-318A in *Third International Symposium on Ceramic Materials and Components for Engines*," ed. by V. J. Tennery (Am. Ceram. Soc., Westerville, OH. 1989).

[19]K. Y Chia and S. K. Lau, "High-Toughness Silicon Carbide," *Ceram. Eng. Sci. Proc.*, **12**, 1845-61 (1991).

[20]N. P. Padture, "In Situ-Toughened Silicon Carbide," *J. Am. Ceram. Soc.*, **77**[2] 519-23 (1994).

[21]K. A. Schwetz, H. Werheit and E. Nold, "Sintered and Monocrystalline Black and Green Silicon Carbide-Chemical Compositions and Optical Properties," *cfi/Ber. DKG*, **80**[12] E37-44 (2003).

[22]ASTM C 1421-99, Standard Test Methods for Determination of Fracture Toughness of Advanced Ceramics at Ambient Temperature, pp. 641-672 in *1999 Annual Book of Standards* (ASTM, Philadelphia, PA 1999).

[23]H. M. Rietveld, "A Profile Refinement Method in Neutron and Magnetic Structures," *J. Appl. Crystallogr.*, **2**, 65-71 (1969).

[24]D. L. Bish and S. A. Howard, "Quantitative Phase Analysis Using the Rietveld Method," *J. Appl. Crystallogr.*, **21**, 86-91 (1988).

[25]E. E. Underwood, *Quantitative Stereology*, (Addison-Wesley, Reading, MA. 1970).

[26]A. Zangvil and R. Ruh, "Phase Relationships in the Silicon Carbide-Aluminum Nitride System," *J. Am. Ceram. Soc.*, **71**[10] 884-90 (1988).

[27]R. M. Williams, B. N. Juterbock, S. S. Shinozaki, C. R. Peters, and T. J. Whalen, "Effects of Sintering Temperatures on the Physical and Crystallographic Properties of β-Silicon Carbide," *Am. Ceram. Soc. Bull.*, **64**[10] 1385-9 (1985).

[28]L. L. Oden and R. A. McCune, "Contribution to the Phase Diagram Al_4C_3-AlN-SiC," *J. Am. Ceram. Soc.*, **73**[6] 1529-33 (1990).

[29]E. L. Pabit, S. Crane, K. Seiben and D. P. Butt, "Grain Boundary and Triple Junction Chemistry of Silicon Carbide with Aluminum and Oxide Additives," this proceedings.

[30]W. Rafaniello, Fabrication and Characterization of Silicon Carbide Alloys: The Silicon Carbide-Aluminum Nitride System (Ph.D. Thesis, University of Utah, 1984).

[31]R. W. Rice, C. C. Wu, and F. Borchelt, "Hardness-Grain Size Relations in Ceramics," *J. Am. Ceram. Soc.*, **77**[10]2539-53 (1994).

CERAMIC-POLYMER COMPOSITES FOR BALLISTIC PROTECTION

Paolo Colombo
Department DICASM, University of Bologna
viale Risogrimento 2, 40136 Bologna, Italy,
Department of Materials Science and Engineering, The Pennsylvania State University
University Park, PA 16802, USA

Eugene Medvedovski
Umicore Indium Products
50 Sims Ave., Providence, RI 02909, USA

Francesco Zordan
Department DICASM, University of Bologna
viale Risogrimento 2, 40136 Bologna, Italy

ABSTRACT

Armor systems require advanced solutions, which rely on a combination of materials with different natures in order to provide adequate ballistic protection and weight saving. Composite layered systems based on monolithic armor ceramic tiles joined with polymer infiltrated ceramic foams have been designed and evaluated, for the first time, for the lightweight ballistic protection. Open cell silicon carbide (SiC) foams of various cell sizes infiltrated with thermosetting or elastomeric polyurethanes were used for this design. The ballistic test results are discussed taking into account the morphological characteristics and the main physical properties of the used materials (facing ceramic tile, ceramic foam and polymers), as well as the test conditions. The proposed design may be considered for structural and vehicular protection solutions, including for blast protection.

INTRODUCTION

Ceramic composites have a variety of applications in diverse engineering areas, where they are subjected to high mechanical and thermal loads. One of the prospective applications of ceramic composites is the ballistic protection structures. The development of stationary and mobile structures against ballistic impacts and explosives, as well as radiation, chemical and biological agents, is an important task in defense applications.

Ceramic-based composites with high toughness are usually reinforced by ceramic or metallic fibers, whiskers or particulates [1, 2], or they may be obtained from the laminated ceramic layers [3]; some of them may be successfully used as armor materials. Another type of composite structures, labeled as interpenetrating composites [4], consists of porous ceramic structures infiltrated with metallic or polymeric materials [5], in which the 3D interconnected ceramic phase affords additional properties (e.g. increased stiffness, wear resistance, etc.) to the infiltrating matrix material.

Ceramic foams may be considered as an important part of composite materials due to their low density, high thermal insulation properties, acceptable strength, etc. as well as unique structure. Among various unique properties of foam ceramics, these materials have a potential as impact energy absorbers due to their cellular structures [6].

A recent study carried out by P. Colombo et al. [7] indeed showed that microcellular ceramic foams, when tested as components of shields protecting spacecrafts from hypervelocity impacts, displayed high performance against high-velocity small projectiles. Some polymeric materials (in bulk or composite forms), which are frequently used as a matrix for the composites, possess adequate mechanical properties and an ability to absorb energy by various mechanisms under mechanical loads [8]. The synergistic combination of positive properties of lightweight ceramic foams and selected polymers may bring positive results in the case of the use of both these materials as components of ballistic protection systems (see below). For instance, rubber encapsulation of cellular ceramics has been shown the significantly increase the amount of impact energy absorbed [9].

Layered composite designs consisting of the materials with different natures are effective for ballistic protection applications if they contain the components with energy dissipation and absorption abilities. The task of the present work was to explore a possibility to use such layered composites consisting of materials with different natures, including solid ceramics and ceramic foam-polymer structures, for ballistic protection. In this case, each constituent will function in accordance with their particular properties.

SELECTED COMPOSITE SYSTEM DESIGN

A layered composite design was selected as potentially effective for the ballistic protection from the performance and manufacturing standpoints. This design included two major components: 1) a ceramic face tile with high hardness and mechanical properties in order to defeat the bullet and to dissipate its impact energy; and 2) a backing material for impact energy absorption and for stopping the bullet and occurred fragments.

As the face component, ballistic grade high-alumina ceramic tiles with high hardness and strength, relatively low brittleness, optimized microstructure and ballistic impact energy dissipation ability were used. Due to high ballistic performance, the alumina tiles with a limited thickness (only 6.0-6.5 mm) were selected. The backing material was made of SiC foams infiltrated by different polymeric materials. A ceramic foam-based system would absorb impact energy by crushing the ceramic foams, and the cellular ceramic structure would interfere with the propagation of the shock wave [10, 11]. Two types of polyurethanes (PU), thermosetting (cross-linked) and elastomeric, were selected for experimental design as the polymeric material. Thermosetting PU is stiff, and it has high compressive strength, while elastomeric PU has a very good damping ability, and it could retard the propagation of shock waves in the system. A recent study conducted by E. Medvedovski et al. indeed showed that the use of some cross-linked polyurethanes in the ceramic-based armor systems increased their ballistic performance. In the case of the polymer-infiltrated ceramic foams, these composites may be considered as ceramic-reinforced polymers possessing significantly increased mechanical properties, especially stiffness and hardness. The proposed design did not have aramid-based (e.g. Kevlar™, Twaron™), laminated polyethylene (Spectra™, Dyneema™) or other backing materials widely used for ballistic protection systems, i.e. the backing consisted solely of a polymer-infiltrated ceramic foam that replaced "traditional" backing. This proposed layered composite(solid ceramic tiles bonded with foam-based 3D-composite backing) may be bonded (attached) to an aluminum or plastic sheet (board), and this "panel" design can be easy installed into any mobile or stationary structure.

EXPERIMENTAL

Materials Selection and Manufacturing

The tiles used as a face component of the armor design were made from alumina ceramics AL98.5 developed and commercially manufactured by Ceramic Protection Corp. [12, 13]. The ceramic material has an Al_2O_3 content of approximately 98.5 wt.-%. It has a uniform fine-crystalline microstructure consisting of corundum grains with a size of 3-5 µm (mostly isometric) bonded by an earth-alkali-aluminosilicate glassy phase. Due to their microcrystalline structure, AL98.5 ceramics possess a high level of physical properties, such as hardness, Young's modulus, sonic velocity, and favorable values of some other characteristics relevant to ballistic performance (Table 1). This ceramic material has a relatively low brittleness (significantly lower than dense carbide-based ceramics) defined by combined values of hardness, fracture toughness and Young's modulus, that may be considered as a positive factor for ballistic performance of ceramics, and optimized high ballistic energy dissipation ability. The ceramic tiles were made by a pressing technology using in-house manufactured spray-dried powders with following firing at relatively low temperature (below 1550°C) without a grinding step after firing (i.e. as-fired tiles were used). Due to the proven high level of ballistic performance of this ceramics, a thickness of only 6-6.5 mm of the tiles was selected for the experimental layered composite designs. The tiles with a format of 100x100 mm were used; however the tiles with other dimensions (with a format up to 150x200 mm and a thickness up to 15 mm) can be manufactured.

As cellular ceramics for subsequent infiltration, open cell SiC foams used as molten metal filters with two different cell sizes (10 and 20 ppi pores-per-inch) were selected. Relative density was approximately 0.13 for both foams. The foams with these pore sizes were chosen due to a possibility to easily infiltrate liquid polymers into these structures without applying external pressure. The foam samples had dimensions of 100x100x20 mm in order to be compatible with a format of the alumina face tiles. The ceramic foams were manufactured by the replica technique, consisted of the coating cellular polymeric sacrificial structures (e.g. reticulated polyurethane with a given pore size distribution) with a ceramic SiC slurry, followed by burning off these performs during firing [14]. It has to be noted that the properties of commercially available molten metal ceramic reticulated filters are generally optimized for permeability and thermal shock resistance, rather than strength, because of their specific application. Foams produced with other technologies [15], affording a different morphology and improved mechanical properties, could also be infiltrated and tested for ballistic protection applications.

The following materials were used for PU preparation. In the case of rigid (thermosetting) PU, isoexter 3446 polyol (Coim Offanengo (CR), Italy) and MDI Suprasec 2085 isocyanate(Huntsman Polyurethanes, Everberg, Belgium) were used as precursors, while Caradol 30-02 polyol (Shell Italia, Milan, Italy) and Suprasec 2444 isocyanate (Huntsman Polyurethanes, Everberg, Belgium) were used for the elastomeric PU preparation. Mixing and handling of precursors were conducted based on the directions from the suppliers of the chemicals.

Infiltration of the porous ceramic foams with polymeric materials was carried out by the vacuum infiltration technique. The ceramic foams were pre-heated at 150°C and then were held under low vacuum (rotary pump, approximately 10^{-2} torr) and then infiltrated with a liquid mixture of the desired PU precursors. The liquids viscosities were of 4000 and 1000 mPa.s at a temperature of 25°C for the thermosetting and elastomeric PU, respectively. Ten minutes setting time at ambient temperature provided proper infiltration and curing of the polymers.

After curing, the infiltrated foams were slightly sanded from the excess of polymer in order to achieve their flat surfaces required for proper bonding with face ceramic tiles. Bonding of the tiles with prepared infiltrated foams was carried out in an autoclave at elevated pressure and temperature using specially selected adhesives and a previously developed procedure for the bonding of armor ceramics with backing materials. The ceramic tiles were preliminary wrapped with prepreg (fiberglass) that is normally used for design and bonding of various armor systems.

Mechanical Testing and Microstructural Analysis

The crushing strength of the foams, both not infiltrated and infiltrated, was measured at room temperature by compression testing, using an Instron 1121 UTM (Instron Danvers, MA) with a cross-head speed of 0.5 mm/min, on samples of a nominal size of 20x20x20 mm, cut from the larger specimens. Each data point represents the average value of 5 to 10 individual tests. Elastic modulus (Young's modulus) was measured using a non-destructive dynamic method based on the impulse excitation of vibration in accordance with ASTM E1876 (Grindosonic, J.W. Lemmens, Leuven, Belgium) on samples of suitable geometry. The ceramic foams' morphology was characterized using optical microscopy and an image analysis software (Image-Pro Plus, Media Cybernetics, Silver Spring, MD).

Ballistic Testing

Ballistic performance of the selected systems was tested using the weapon M16 with the ammunitions 5.56x45-mm SS109 with a steel tip ball, 7.62x51-mm NATO Ball Full Metal Jacket (FMJ) with a lead core and 7.62x63-mm Armor Piercing (AP) M2 FMJ with a tungsten carbide core. Depending on the ammunition, the bullet weight, velocity and energy were varied. The projectile velocity was controlled using a chronograph. The projectile velocities were 990-1000, 840-890 and 845-870 m/s, respectively, i.e. the velocity values were in accordance with the values normally used for ballistic testing and recommended by the NIJ Standards. The distance between the weapon and the target was 12 m. The trauma after shooting was measured using a Roma Plastilina modeling clay supporting armor systems on the back; the trauma in clay duplicated the trauma in the armor. A trauma pack (a block of Kevlar simulated armor vest) was placed behind the clay in order to capture a bullet in the case of penetration. The damage zone of the ceramics and the backing, including components fragmentation, and the bullets after shooting were observed.

RESULTS AND DISCUSSION

Microstructure study of the un-infiltrated SiC foams (see Fig. 1a,b) showed completely open interconnected porosity comprised of a web of ceramic hollow struts. The hole in the struts derives from the burn-off the polymeric sacrificial template during high temperature firing of the SiC powder coating. This observed structure had a similar character for both studied SiC foams (with 10 and 20 ppi). Morphological analysis of the infiltrated foams revealed that it was possible to completely infiltrate the ceramic foam structures with both elastomeric and thermosetting PU precursor solutions (see Fig. 2a,b). However, due to the higher viscosity of the thermosetting PU precursor solution, some trapped air was retained in the PU phase. It was worth observing that in some cases, the space inside the hollow struts was also partially infiltrated by the polymeric phase, as shown in Fig. 2b. In this paper, no specific investigation was performed on the nature of the polymer-ceramic interface.

Ceramic foams and foams infiltrated with selected polymers were mechanically tested in order to compare the influence of a foam macrostructure and a type of polymer on mechanical properties of the interpenetrating composites and to interpret their behavior at the ballistic testing. The mechanical test results are reported in Table 2. As-received ceramic foams displayed limited strength, typical of cellular ceramics produced by the replica technique. Mechanical properties and bulk density were slightly higher for the foams with porosity of 10 ppi comparatively with the foams of 20 ppi. It may be explained by the greater cell wall thickness for the former ceramic foams. The foams infiltrated with polyurethanes varied their strength according to the type of PU used. While the composite material produced using elastomeric PU did not display a crushing strength significantly different from that of the un-infiltrated ceramic cellular structures, using thermosetting PU resulted in a significant increase of compression strength of the foams. The difference in behavior of infiltrated foams under mechanical loads is attributable to the variation in strength of the PU phase itself (see Table 2). An increment in stiffness (that may be expressed by the values of modulus of elasticity) was observed for the composites infiltrated with thermosetting PU, but the data could not be obtained for the foams infiltrated with elastomeric PU due to its damping effect on the vibrations generated during the measurement.

When an un-infiltrated ceramic foam is subjected to a compressive load, it deforms elastically up to a maximum load (usually occurring at strains of about 5%), after which brittle crushing occurs (with more or less sudden drop of stress), and deformation follows a plateau region before increasing rapidly due to densification of the crushed debris [16] (Fig. 3). Typically, after reaching the peak load, the ceramic foam is subjected to macroscopic cracking along the direction of the applied stress, and if the loading ram is lifted, the tested sample falls apart in several large fragments [17] (Fig. 4a). The use of the thermosetting PU phase resulted in the modifying of the behavior of the ceramic foams under compression. The stress-strain curves for infiltrated ceramic foams did not show a drop in the stress after reaching the peak load (Fig. 3), and the samples remained in one piece (or fragmented in a very limited amount of pieces) after the removal of the applied load. This behavior may be explained by the binding effect of the interconnected polymeric phase (see Fig. 4b as an example). Cracking phenomena during the loading seemed to occur mainly in the brittle ceramic phase (Fig. 4c). Infiltration with thermosetting PU afforded to the composite structure much higher energy absorption, i.e. work of fracture (see the area underneath the stress-strain curve of Fig. 4) in comparison to the un-infiltrated foam. Due to low strength and low Young's modulus, elastomeric PU infiltrating phase did not influence much the stress-strain behavior of the interpenetrating polymer-ceramic composite, but it helped maintaining the tested sample into a coherent piece until large deformations (greater than 0.5) were reached.

A strong influence of the design and mechanical properties of the composite components and, especially, of the selected polymer nature, on ballistic performance was observed. Preliminary testing of the system based on the SiC foam infiltrated by thermosetting PU but without a face alumina tile showed non-satisfactory performance against the 7.62x51-mm NATO projectile (the testing of the infiltrated foam using an armor piercing projectile, i.e. with harder core and with higher energy, was not undertaken). In the case of SiC foams infiltrated with thermosetting PU, the 5.56x45-mm SS109 and 7.62x51-mm NATO rounds were stopped in the backing (Fig. 5a,b), while the 7.62x63-mm AP round was stopped only in the first layers of the trauma pack (Fig. 6a,b). A similar system made with elastomeric PU could not defeat even the 7.62x51-mm NATO round and, especially AP round (Fig. 7a,b). The difference in ballistic protection of the composites made with two different types of PU is in a good correlation with their mechanical properties.

For the face ceramic tile, the observed fragmentation and crack formation from ballistic impacts of the designed composites had a similar character as for the high-alumina ceramics bonded to Kevlar backing (described previously [12, 13]). The ceramic tile zone close to the impact point consists of fragments of various sizes to fine powder with a locus of conoidal coaxial cracks with a presence of radial tensile and spall cracks. The backing materials consisted of SiC foams (both 10 and 20 ppi) infiltrated with thermosetting PU also fragmented in chunks of different sizes; the powder formation was minimal. The reinforcement of thermosetting PU achieved by the use of the SiC foams promoted a large chunks formation upon impact and a decrease of "elasticity" of PU. The destruction of the composites and occurred trauma were significantly lower in the case of the 5.56x45-mm SS109 projectile due to its lower energy comparatively with NATO Ball and AP projectiles. A comparison of ballistic performance of the armor systems made using SiC foams with larger or smaller pore sizes (10 or 20 ppi) did not indicate any significant difference. This is probably due to the limited difference in strength and elastic modulus between the two ceramic foams. In the case of elastomeric PU, a fragments formation practically was not observed in the backing, and a trauma was minimal due to penetration. These results confirm the idea that the use of an infiltrating polymeric material with significantly higher strength is required for providing adequate ballistic protection capability. Also, in this case, the difference between the composites made from the SiC foams with different pore sizes was not found.

CONCLUSIONS AND FUTURE WORKS

The layer design consisted of armor ceramic tiles bonded to ceramic foams infiltrated using two types of polyurethanes was evaluated, for the first time, as armor system. Fully infiltrated, 3D interpenetrating ceramic-polymer composites were manufactured using vacuum-assisted processing. The composites consisting of SiC foams and thermosetting (cross-linked) PU demonstrated a relatively high level of mechanical properties. The armor systems based on only 6-mm alumina armor tiles bonded with SiC-thermosetting PU composite backing could defeat 5.56x45-mm SS109 and 7.62x51-mm NATO Ball FMJ projectiles; and these systems may have a potential for ballistic and blast protection for vehicular and structural applications. The use of elastomeric PU as the infiltrating phase did not afford sufficient properties to the composite to make it of interest for ballistic protection applications.

Obtained results of initial testing conducted at the present work allow exploring further a new approach of the design of armor systems with lower weight. The use of the backing systems based on ceramic foams infiltrated with a proper PU (or some other materials) allow replacing expensive polyaramid or polyethylene backing and a labor intensive lamination process.

The following directions of future works may be considered:

- Design optimization, including optimization of the ratio of thicknesses of a face ceramic tile and a polymer-infiltrated ceramic foam; for example, a slight increase of a ceramic tile thickness (e.g. up to 7-8 mm) may significantly increase ballistic performance against armor-piercing projectiles;
- Optimization of ceramic foam structure (pore size) and strength; the use of foams with significantly smaller pore sizes may change ballistic fracturing behavior and improve ballistic performance, as well the use of foams possessing improved mechanical properties produced by other processing techniques;
- Search of proper polymer materials with optimized high stiffness and elasticity and with acceptable workability (i.e. with a good ability to fill the ceramic foam and a good wetting of the ceramic foam surface) that may improve ballistic performance and decrease weight of the armor system;
- Optimization of bonding of face ceramic tiles with infiltrated ceramic foams;
- Wide evaluation of the optimized designs in terms of ballistic performance, especially against AP projectiles, as well as evaluation in radiation and harsh chemical environments.

ACKNOLEDGEMENTS

Foseco, USA, is greatly acknowledged for providing the SiC ceramic foam samples.

REFERENCES

[1] W. Krenkel, "Carbon Fiber Reinforced CMC for High-Performance Structures", *International Journal of Applied Ceramic Technology*, 1 [2] (2004)188-200.

[2] D.J. Viechnicki, M.J. Slavin, M.I. Kliman, "Development and Current Status of Armor Ceramics", *American Ceramic Society Bulletin*, 70 [6] (1991) 1035-1039.

[3] N. Orlovskaya, M. Lugovy, V. Subbotin, et al., "Design and Manufacturing B_4C-SiC Layered Ceramics for Armor Applications"; pp 59-70 in Ceramic Transactions, Vol. 151, *Ceramic Armor and Armor Systems*. Edited by E. Medvedovski, American Ceramic Society, Westerville, OH, 2003.

[4]. D.R. Clarke, Interpenetrating Phase Composites", *Journal of The American Ceramic Society*, 75 [4] (1992) 739-758.

[5]. H.X. Peng, Z. Fan, J.R.G. Evans, "Bi-Continuous Metal Matrix Composites", *Materials Science and Engineering,* A303 (2001) 37-45.

[6]. L.J. Gibson and M.F. Ashby. "Cellular Solids, Structure and Properties", 2nd edition. Cambridge University, 1999.

[7] P. Colombo, A. Arcaro, A. Francesconi, et al., "Effect of Hypervelocity Impact on Microcellular Ceramic Foams from a Preceramic Polymer", *Advanced Engineering Materials*, 5 [11] (2003) 802-805.

[8]. G.C. Jacobs, J.F. Fellers, S. Simunovic, et al., Energy Absorption in Polymer Composites for Automotive Crashworthiness", *Journal of Composite Materials*, 36 [7] (2002) 813-850.

[9]. V. Jain, R. Johnson, I. Ganesh, et al., "Effect of Rubber Encapsulation on the Comparative Mechanical Behavior of Ceramic Honeycomb and Foam", *Materials Science and Engineering*, A347 (2003) 109-122.

[10]. D.W. Dixon-Hardy and M. Dwomoh, "The Attenuation of Shock Waves by Ceramic Foams"; pp 1-15 in *European Congress of Computational Methods in Applied Sciences and Engineering*, Barcelona, Spain, 2000.

[11]. A. Levy, G. Ben-Dor, B.W. Skews, et al., Head-om Collision of Normal Shock Waves with Rigid Porous Materials", *Experimental Fluids*, 15 (1993) 183-190.

[12] E. Medvedovski, "Alumina Ceramics for Ballistic Protection", *American Ceramic Society Bulletin*, **81** [3] (2002) 27-32, [4] (2002) 45-50.

[13] E. Medvedovski, "Armor Alumina Ceramics"; pp 91-101 in Ceramic Transactions, Vol.134, *Ceramic Armor Materials by Design*. Edited by J.W. McCauley, A.Crowson, W.A. Gooch, Jr., et al., American Ceramic Society, Westerville, OH, 2002.

[14]. K. Schwartzwalder and A.V. Somers, "Method of Making Porous Ceramic Articles", U.S. Patent 3,090,094 (1963).

[15]. P. Sepulveda and J.G.P. Binner, "Processing of Cellular Ceramics by Foaming and in situ Polymerization of Organic Monomers", *Journal of the European Ceramic Society*, 19 (1999) 2059-2066.

[16]. M.F. Ashby, "The Mechanical Properties of Cellular Solids", *Metall. Transactions*, 14A (1983) 1755-1769.

[17]. Y. Yamada, K. Shimojima, M. Mabuchi, et al., "Compressive Deformation Behavior of Al2O3 Foam", *Materials Science and Engineering*, A277 (2000) 213-217.

Table 1. Physical Properties of Alumina Ceramics AL98.5 (Pressed)

Density	3.82 – 3.84 g/cm^3
Open Porosity	less than 0.1%
Vickers hardness *Hv10*	1320 – 1400 kg/mm^2 (12.9 - 13.7 GPa)
Fracture toughness K_{Ic}*	3.0 – 3.3 MPa.m$^{1/2}$
Young's modulus *E*	330 – 400 GPa
Sonic velocity *c*	10200 – 10800 m/s
Flexural strength	260 – 280 MPa
Brittleness factor *B***	(420 – 460).10^6 1/m
Ballistic energy dissipation ability *D****	(1.8 – 1.95).10^{12} 1/s

*Fracture toughness was determined by indentation technique
**Brittleness was calculated using a formula $B = Hv.E/K_{Ic}^2$
***Ballistic energy dissipation was calculated using a formula
$D = 0.36.Hv.E.c/ K_{Ic}^2$

Table 2. Properties of As-Received and Infiltrated Ceramic Foams

SiC foams	Average cell size, mm	Bulk density, g/cm^3	Compression strength, MPa	Elastic modulus, GPa
10 ppi, un-infiltrated	4.46±1.17	0.437±0.056	1.1±0.3	1.80±0.14
20 ppi, un-infiltrated	3.09±0.61	0.402±0.024	0.8±0.1	1.74±0.10
10 ppi, thermosetting PU	4.46±1.17	1.093±0.198	71.5±4.5	3.69±0.30
20 ppi, thermosetting PU	3.09±0.61	1.180±0.049	42.0±1.5	7.11±1.71
10 ppi, elastomeric PU	4.46±1.17	1.023±0.025	1.0±0.2	n.d.*
20 ppi, elastomeric PU	3.09±0.61	1.044±0.052	1.1±0.1	n.d.*
Elastomeric PU	-	0.942±0.098	n.d.*	n.d.*
Thermosetting PU	-	0.962±0.129	64.7±2.1	1.79±0.49

* not determined due to the damping of vibration by elastomeric PU phase

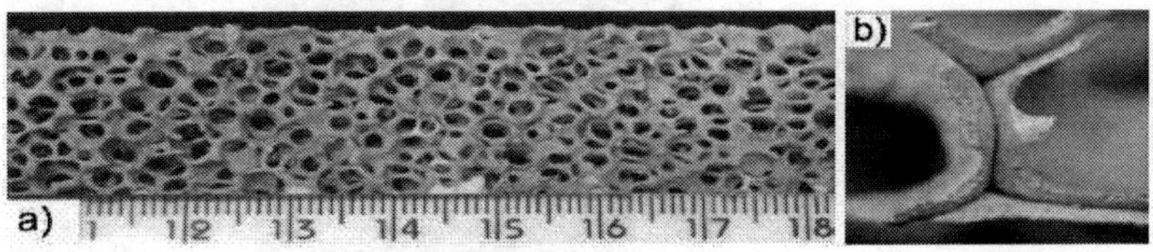

Fig.1. Un-infiltrated SiC foam (20 ppi)
a) general view; b) detail of a hollow strut

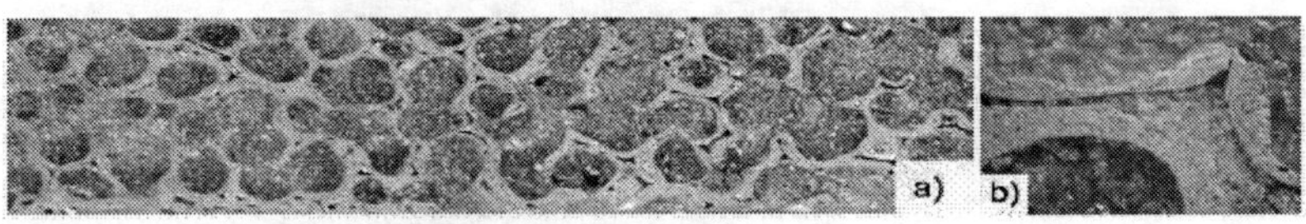

Fig. 2. SiC foam infiltrated with thermosetting PU
a) general view; b) detail of a strut

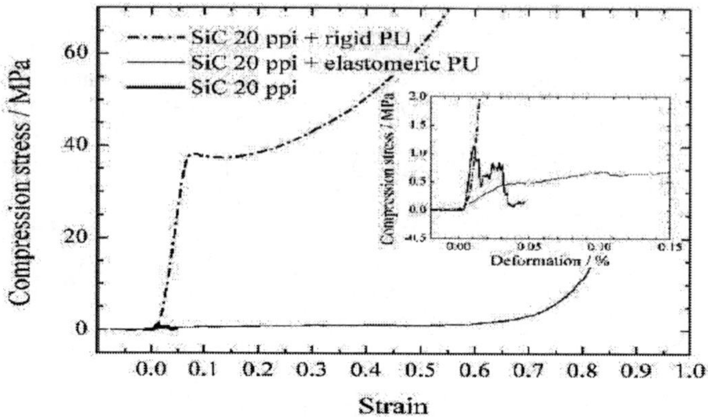

Fig. 3. Stress-strain curves for SiC foams (20 ppi), as-received, infiltrated with elastomeric and thermosetting PU

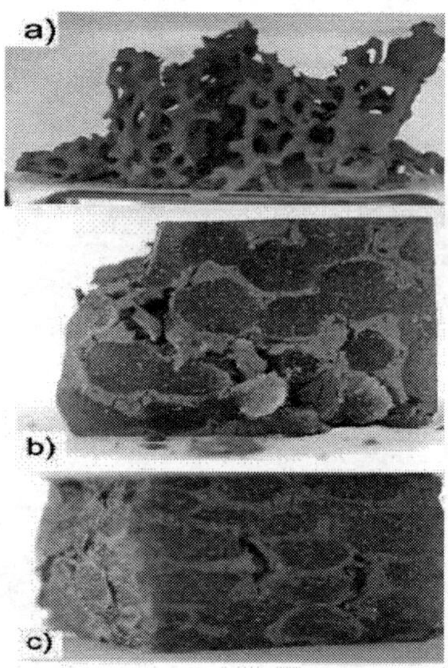

Fig. 4. SiC foams (20 ppi) after compression

a) un-infiltrated (test stopped at a relative deformation of 0.3%)
b) infiltrated with thermosetting PU (test stopped at a relative deformation of 0.3%)
c) infiltrated with thermosetting PU (test stopped at a relative deformation of 0.14%); crack formation in the SiC phase

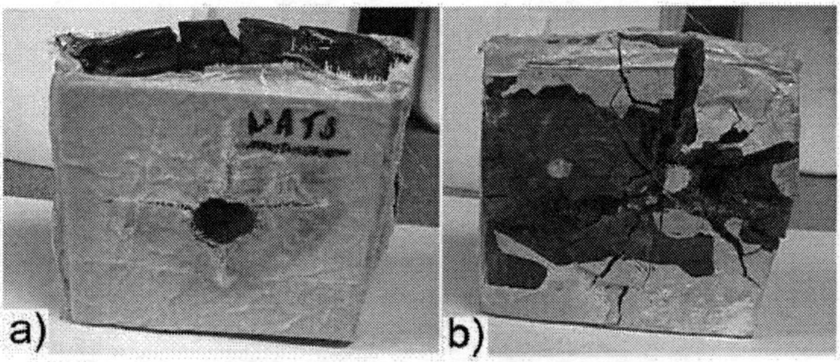

Fig.5. Armor system with SiC foam (20 ppi) infiltrated with thermosetting PU after ballistic testing with 7.62x51-mm NATO Ball FMJ
a) face view; b) back view

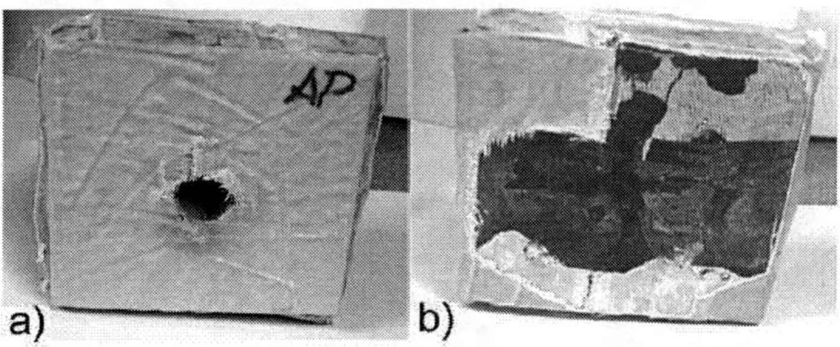

Fig. 6. Armor system with SiC foam (20 ppi) infiltrated with thermosetting PU after ballistic testing with 7.62x63-mm AP M2
a) face view; b) back view

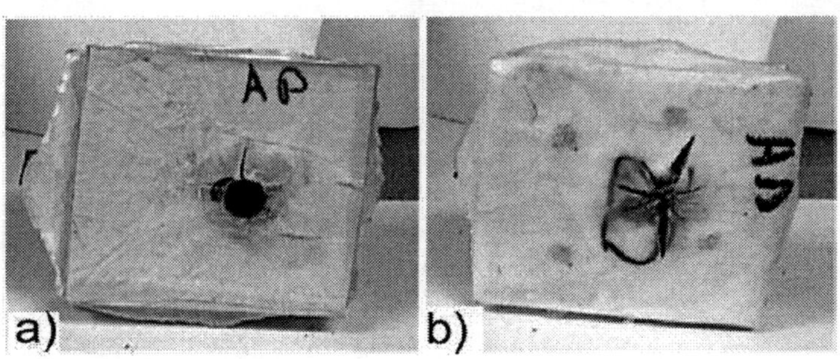

Fig. 7. Armor system with SiC foam (20 ppi) infiltrated with elastomeric PU after ballistic testing with 7.62x63-mm AP M2
a) face view; b) back view

GRAIN BOUNDARY AND TRIPLE JUNCTION CHEMISTRY OF SILICON CARBIDE WITH ALUMINUM OR ALUMINUM NITRIDE ADDITIVE

Edgardo Pabit, Samantha Crane, Kerry Seiben and Darryl P. Butt
University of Florida
Materials Science and Engineering Department
P.O. Box 116400
Gainesville, FL 32611

Darin Ray, R. Marc Flinders, and Raymond A. Cutler
Ceramatec, Inc.
2425 South 900 West
Salt Lake City, UT 84119

ABSTRACT

Grain boundary and triple grain junction compositions of SiC sintered with 1.65 wt. % Al or 2.5 wt. % AlN (with or without B_4C and C) were investigated through the use of transmission electron microscopy (TEM) and energy dispersive x-ray spectroscopy (EDS). Segregation of Al and O in triple grain junctions occurred for all Al-containing materials. The only material with clean grain boundaries was processed with no Al using B_4C and C as densification aids. The presence of aluminum and oxygen at triple points does not insure intergranular fracture. The addition of B_4C and C to Al containing materials did not reduce the O concentration, but allowed transgranular fracture due to the absence of O along grain boundaries.

INTRODUCTION

Silicon carbide is the armor material of choice for medium to heavy threats due to its lightweight and improved ballistic response compared to boron carbide. SiC is prepared by hot pressing or pressureless sintering. Typical densification aids are B (or B_4C) and C[1], Al, or AlN.[2,3] Hot pressed SiC using AlN as a densification aid[4] is the state-of-the-art armor material sold under the trade name of SiC-N by Cercom. Early work by Alliegro, et al.[5] showed that Al was an effective densification aid. Al was combined with C[6], or with B and C[7], in order to densify SiC. Higher additions of Al, in combination with B and C, result in high-toughness materials[8] due to the combination of intergranular fracture and elongated grains.[9]

Transmission electron microscopy has been used to study SiC densified by different means. When only boron and carbon are used as densification aids, grain boundaries are clean and boron carbide precipitates are observed.[10] When Al is introduced, grain boundaries are enriched with aluminum[11] and oxygen[12], with glassy grain boundaries commonly observed.[13] Larger additions of Al in combination with B and C result in a variety of Al-containing materials at triple points including Al_2O_3, $Al_8B_4C_7$, Al_4C_3, Al_2OC and mullite.[14],[15] AlN additions limit grain size but generally result in clean grain boundaries when carbon is present to remove oxygen.[16] AlN and Al_2O_3 can be added simultaneously to control grain size and polytypes present.[17] When Al_2O_3 is present, Al and O can segregate to grain boundaries[18] resulting in a grain boundary with a thickness of less than 1 nm.[19]

Due to the ability to control fracture mode from transgranular to intergranular by the judicious selection of sintering additive, amount of additive, and processing conditions[20] it is of interest to compare the ballistic performance of difference SiC materials. In this work, the amount of aluminum, added as Al or AlN, was kept constant. B_4C and C were used as additional additives with the belief that they would reduce the oxygen content in the starting α-SiC powder. The purpose of this study was to compare the grain boundaries and triple points of SiC processed with Al or AlN (with and without B and C) to that of SiC-N and solid state densified SiC.

EXPERIMENTAL PROCEDURE

Materials

Table I lists the materials studied. SiC-N is a commercially available SiC from Cercom, Inc. produced via the pressure assisted densification method. The remaining materials were produced using grade UF-15 α-SiC from H. C. Starck. The additives used to densify the SiC materials were C (CRC, grade 720 Resin), B_4C (H. C. Starck, grade B_4C-HS), AlN (Tokuyama Soda, grade F) and Al (Valimet, grade H-3). The 1.65 wt. % Al level was selected to be equi-molar in Al to the 2.5 wt. % AlN. All compositions were prepared by batching 600 grams of powder in two-liter high-density polyethylene (HDPE) jars filled with 1.6 kg solid state sintered SiC media and 700 g reagent grade acetone. The slurries were mixed for 16 hours on a ball mill in order to disperse the agglomerates and the powders were stir dried before screening through an 80-mesh screen. Powder compositions batched with carbon resin were pyrolyzed by heating in N_2 to 600°C. Billets (45 mm x 45 mm x 6 mm) were hot pressed at 28 MPa inside graphite dies by heating to 1500°C in a vacuum of 1 torr, holding at 1500°C for one hour to remove SiO and CO from the samples, and then backfilling with Ar and heating to 2100°C for one hour. The sample designation, hot pressing condition, starting material composition and the resulting polytypes of the fabricated ceramics are shown in Table I.

Characterization

All microhardness data were obtained using a Vickers indenter (Leco Model LM-100) under one kilogram load (HV1) on polished surfaces. The dwell time at load was nominally 15 seconds. Rietveld analysis was used to determine SiC polytypes present in the densified samples with X-ray diffraction patterns collected on the ground surface of a bar, from a diffraction angle of 30-80°, with a step size of 0.02°/step and a counting time of 4 sec/step.

A high resolution transmission electron microscope (model JEOL 2010F) was operated at 200kV. The beam size used for the scanning transmission electron microscopy and energy dispersive x-ray spectroscopy was 0.8 nm. Oxygen and nitrogen were analyzed on densified

Table I. Starting Composition, Processing Condition and Phase Assemblage of SiC-Based Ceramics.

Sample Name	Composition (wt.%)	Processing Condition	Phase Assemblage (wt. %)			
			3C	4H	6H	15R
SiC-N	Proprietary	Proprietary	0.0	0.0	100.0	0.0
SiC(Al)	1.65 Al	2100°C/1 hr	0.0	61.6	33	5.4
SiC(Al-B-C)	1.65 Al – 0.5 B_4C – 2C	2100°C/1 hr	0.0	21.2	71.7	7.1
SiC(AlN)	2.5 AlN	2100°C/1 hr	3.3	26.8	62.2	7.7
SiC(AlN-B)	2.5 AlN – 0.5 B_4C	2100°C/1 hr	0.0	52.3	44	3.7
SiC(AlN-B-C)	2.5 AlN – 0.5 B_4C – 2C	2100°C/1 hr	0.0	49.4	45.7	4.9
SiC(B-C)	0.5 B_4C – 2 C	2100°C/1 hr	0.0	0.0	94.8	5.2

samples using ASTM method E1409 for the determination of oxygen by the inert gas fusion/thermal conductivity detection technique with a Leco TC-136 model analyzer

RESULTS AND DISCUSSION

Characterization

All of the fabricated materials sintered to densities greater than 3.16 g/cc but the data in Table II make it clear that C additions did not aid Al-additives in densification. The addition of boron(as B_4C), which promotes grain growth[21], did not have a deleterious effect on densification. Results of the Rietveld analysis (see Table I) performed on the materials fabricated showed that there is little evidence for complete solid solution formation with AlN. Based on the work of Zangvil and Ruh[22] one could expect a 4H solid solution to occur between SiC and AlN. The starting powder was primarily 6H, and since there is substantial 6H content in all of the materials containing nitrogen a complete solid solution did not form. No 2H SiC was detected since the starting powder was 6H and the additive levels were low. The addition of Al promoted the transformation to 4H SiC[23] with less of this polytype present when B_4C and C were added.

Table II shows the mechanical properties of the SiC materials studied. The Vickers hardness ranges from 20 to 26 GPa, while the SEPB toughness was observed to vary from 2.4 to 5.3 MPa-m$^{1/2}$. High toughness in SiC ceramics is generally dependent on promoting some intergranular fracture so that crack bridging can occur.[24] While polytype formation may be an indicator of elongated grain growth, it is not a predictor of toughness.[25] Toughness in the ABC system is independent of starting polytypes but very dependent on amount of additive.[9,24] The amount of Al in the SiC(Al-B-C) sample was too low to promote intergranular fracture despite exceeding the solubility limit. Hardness is dependent on the additive system and the fracture mode.[24,25] The highest hardness value was observed for SiC with B_4C/C additives. This material also shows the lowest toughness value due to its transgranular fracture mode. As shown in Figure 1, the fracture mode can be discerned by viewing crack paths from Vickers indents as well as fracture surfaces. The SiC material with comparable hardness and toughness to that of the commercially available SiC-N was the SiC processed only with Al additive. This material has elongated grains in combination with a mixed fracture mode. A high degree of control is possible by varying additives, additive level and processing conditions.[20] The SiC(AlN) was lower in toughness than SiC(Al) due to a finer grain size.[20] Adding B_4C and C increased the grain size, promoting transgranular fracture and decreasing toughness.[20]

Table II. Properties, Chemical Analysis, and Mode of Failure of the SiC Materials.

Sample Name	Density (g/cc)	Chemical Analysis (wt. %)		Hardness (HV1, GPa)	Toughness (MPa-m$^{1/2}$)	Mode of Fracture
		Oxygen	Nitrogen			
SiC-N	3.22	0.56	0.17	21.9±0.3	5.3±0.1	Intergranular
SiC(Al)	3.21	0.76	0.01	22.1±0.7	4.7 ± 0.4	Mixed
SiC(Al-B-C)	3.16	0.98	0.01	20.3±0.3	3.1 ± 0.1	Transgranular
SiC(AlN)	3.21	0.44	0.20	21.9±0.5	3.5 ± 0.2	Intergranular
SiC(AlN-B)	3.20	0.38*	0.84*	21.0±0.6	3.4 ± 0.4	Mixed
SiC(AlN-B-C)	3.16	0.72	0.59	22.7±0.3	2.8 ± 0.2	Transgranular
SiC(B-C)	3.19	0.02	0.01	25.7±0.4	2.4 ± 0.1	Transgranular

*Obtained on samples densified at 2200°C.[20]

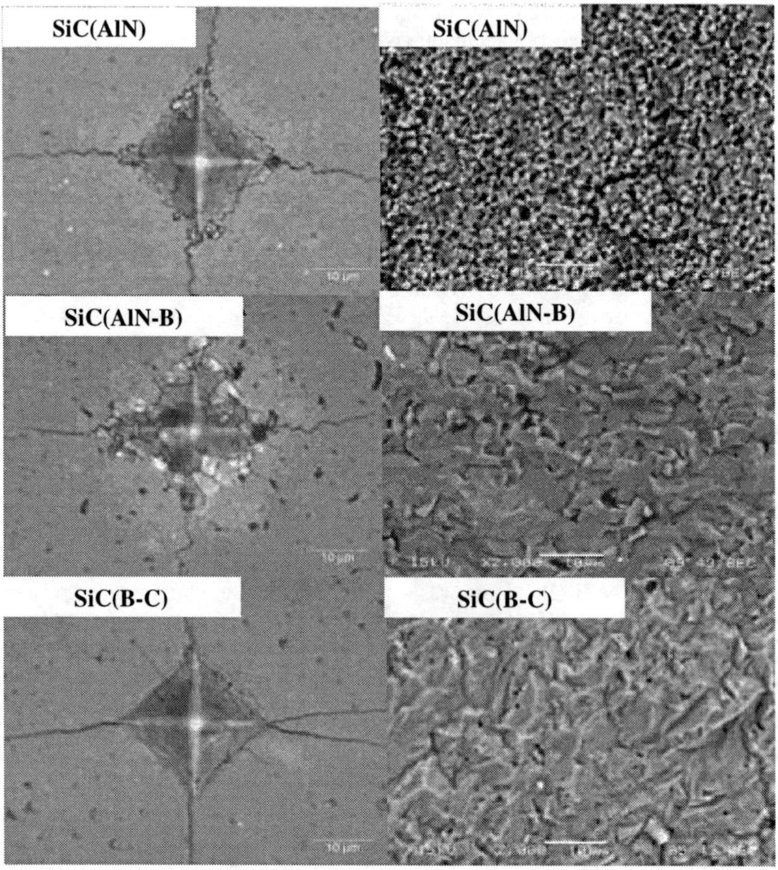

Figure 1. Vickers hardness indents (left) and fracture surfaces (right) for selected samples showing how fracture mode listed in Table II was discerned. All markers are 10 μm.

The chemical analysis in Table II shows that the only material with low oxygen content is the SiC processed without Al or AlN. The 1500°C hold in vacuum was apparently insufficient to remove all of the oxygen from these samples. In addition to both Al and AlN starting materials containing oxygen[20], they also promote a more rapid densification. The removal of O from SiC is via SiO, Al_2O, and CO[26] volatility would therefore be reduced, as porosity was closed up sooner in the sintering cycle. The addition of B_4C and C likely further increased the amount of retained O and N by promoting earlier densification.

TEM and EDS Results

Figure 2 shows bright field images of the seven materials investigated. Residual carbon was present in materials to which a phenolic resin was added. Stacking faults are obvious in all

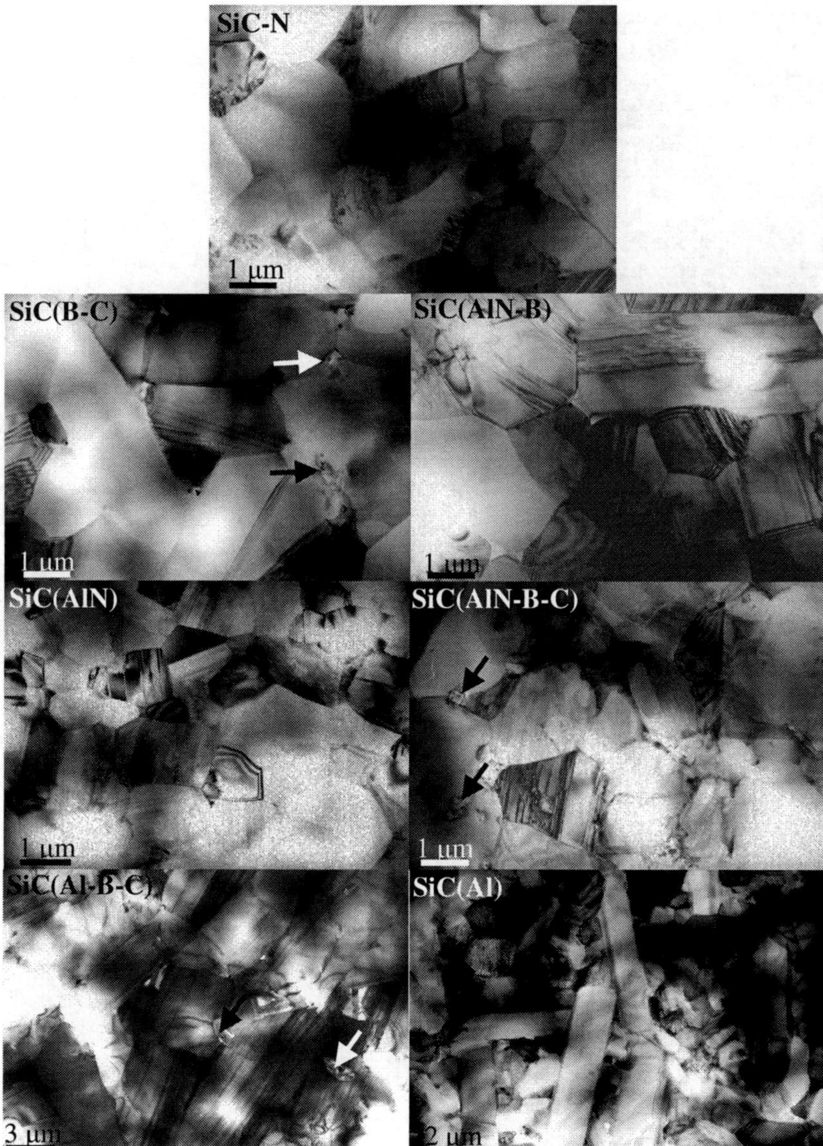

Figure 2. Bright field TEM images of the SiC materials studied. Arrows shows the carbon grains observed from the three samples showing either a transgranular or mixed mode of failure.

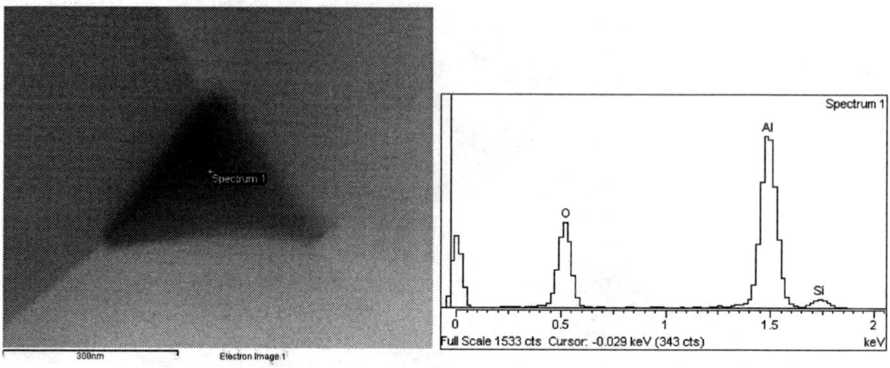

Figure 3. TEM image (left) and EDS spectrum (right) of SiC-N show that triple point is filled with Al and O. The marker on the TEM image is 300 nm.

materials. Grain size for all of the materials is in the 1-5 μm range.[20] Al and O were observed in the triple points of all materials except the SiC densified with only B_4C and C. Metallic impurities were found in all materials due to transition metals in the starting SiC powder.

Figure 3 shows an example of an aluminum oxide-based liquid that solidified in the SiC-N material. The higher thermal expansion of the aluminum oxide relative to the silicon carbide creates a residual tensile stress field which promotes intergranular fracture. However, not all triple points contain similar amount of aluminum and oxygen, as can be seen in Figure 4. This triple point also contains some Al and O similar to what was observed along grain boundaries in this material. Figure 5 indicates that the grains contain much lower amounts of Al and O than the grain boundaries and triple points. While N is present in the material it does not segregate at

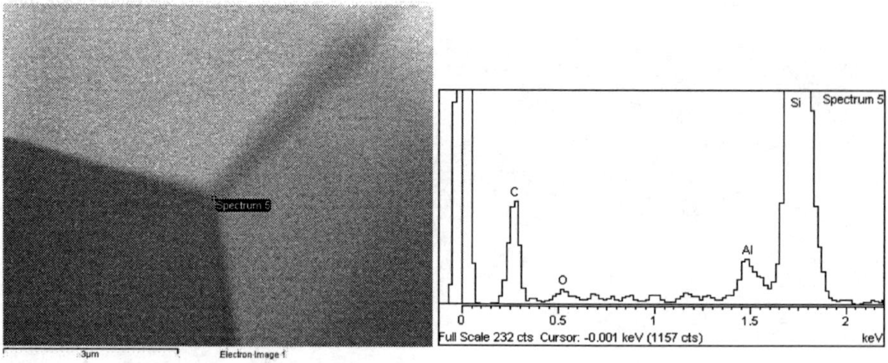

Figure 4. TEM image of SiC-N showing a three-grain junction containing Al and O (left) with corresponding EDS spectrum (right). Marker is 3 μm.

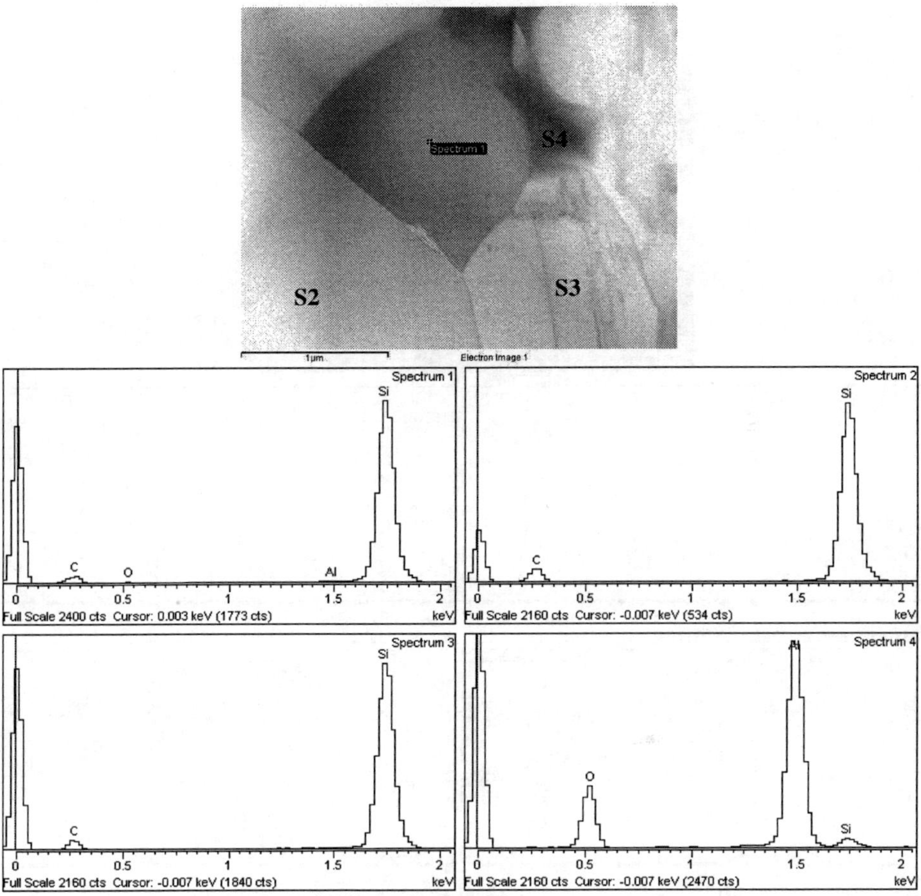

Figure 5. TEM image (top) and four EDS images of SiC-N. Spectrum 1-3 are inside SiC grains showing low O and Al, while spectrum 4 shows a material high in Al and O. Marker is 1 μm.

triple points or grain boundaries. As is clearly shown, the Al and O contents are undetected by EDS within the grains compared to that at the triple point or grain boundaries.

In contrast, Figure 6 shows that the grain boundaries in SiC processed with only B_4C and C are free of oxygen, in accordance with the chemical analysis shown in Table II. Apparently the free carbon in the material was able to remove all of the SiO_2 on the surface of the SiC particles. This material fails primarily transgranularly since there is no grain boundary film that provides an easier fracture path.

When both Al and C are present, as was the case for the AlN-B_4C-C, there is still Al and O within triple points as shown in Figure 7. Based on the chemical analysis (see Table II), there is more O and N present with than without C. There is no evidence, however, of nitrogen

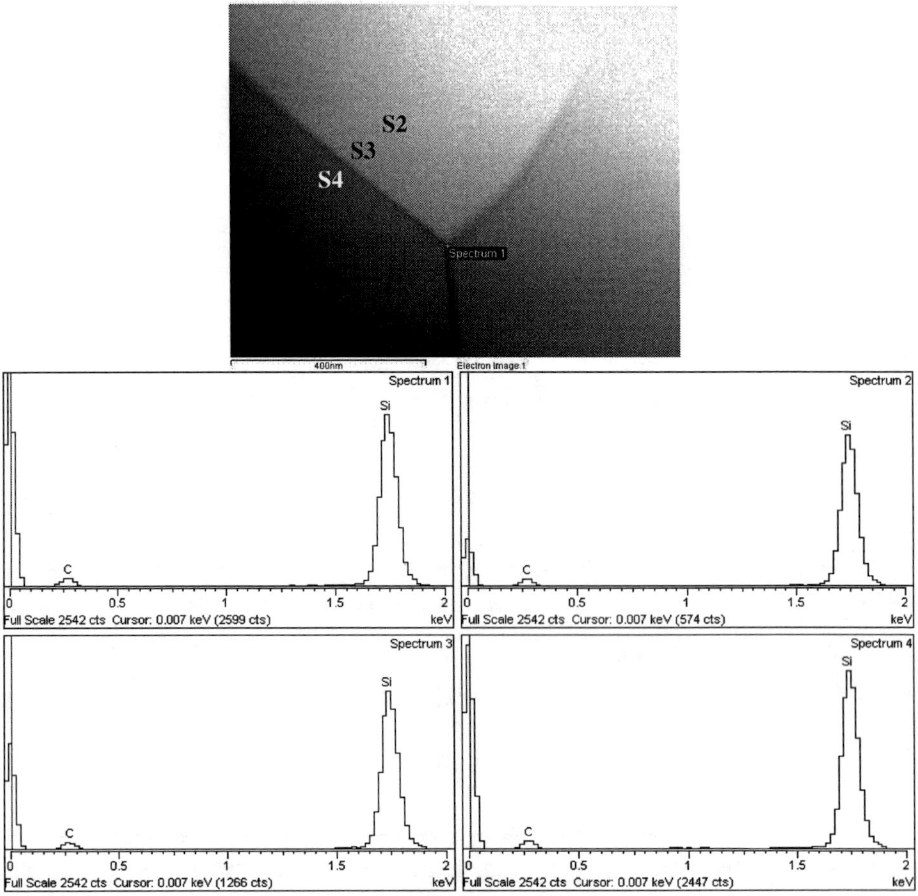

Figure 6. TEM image (top) and four EDS spectra for SiC(B-C). Spectrum 1 is at three-grain junction, while spectrums 2-4 are within the grains. All EDS spectra are similar showing no oxygen present, in accord with chemical analysis. Marker is 0.4 μm.

segregation at triple points or grain boundaries. The majority of the oxygen tends to collect in triple points and is associated with aluminum. While some grain boundaries are clean, others have some Al associated with them. There is no evidence for O along grain boundaries, which appears to lead to transgranular instead of intergranular fracture. This leads to low toughness, similar to SiC sintered with just B_4C and C, but with lower hardness due to the incorporation of Al into the grains. It is clear that the presence of Al and O at triple points does not necessitate an intergranular or mixed mode fracture.

Figure 8 shows that when Al is used instead of AlN that grain boundaries contain both Al and O, although in lesser concentration than at triple points. This permits a mixed fracture mode,

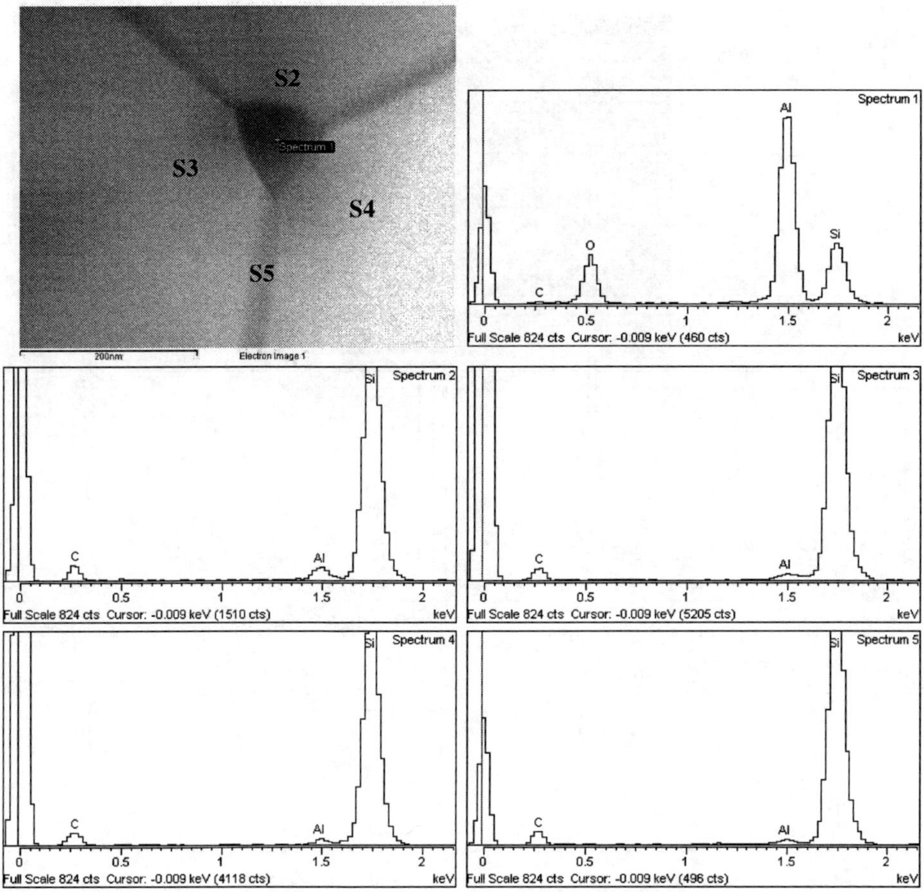

Figure 7. TEM image (top left) and five EDS spectra for SiC(AlN-B-C). Spectrum 1 is at triple point, spectrums 2-4 are within the grains, and spectrum 5 is at a grain boundary. Al and O are present in the triple point, but there is no evidence for N segregation to grain boundaries or triple points. Marker is 0.2 μm.

which appears to be independent of grain size in this material.[20] This leads to a hardness-toughness trade-off with grain size.[20] When B_4C and C are added, the oxygen concentrates in the triple points leading to transgranular fracture. Table III summarizes the observations for all of the materials. From these limited observations it appears that the triple point chemistry is less important than the grain boundary chemistry. When carbon is added, it apparently changes the dihedral angle resulting in less oxygen along grain boundaries. Clearly more work is required to understand more fully what controls the fracture mode in these materials. Quantification of the fracture mode and additional TEM studies are necessary for this to occur.

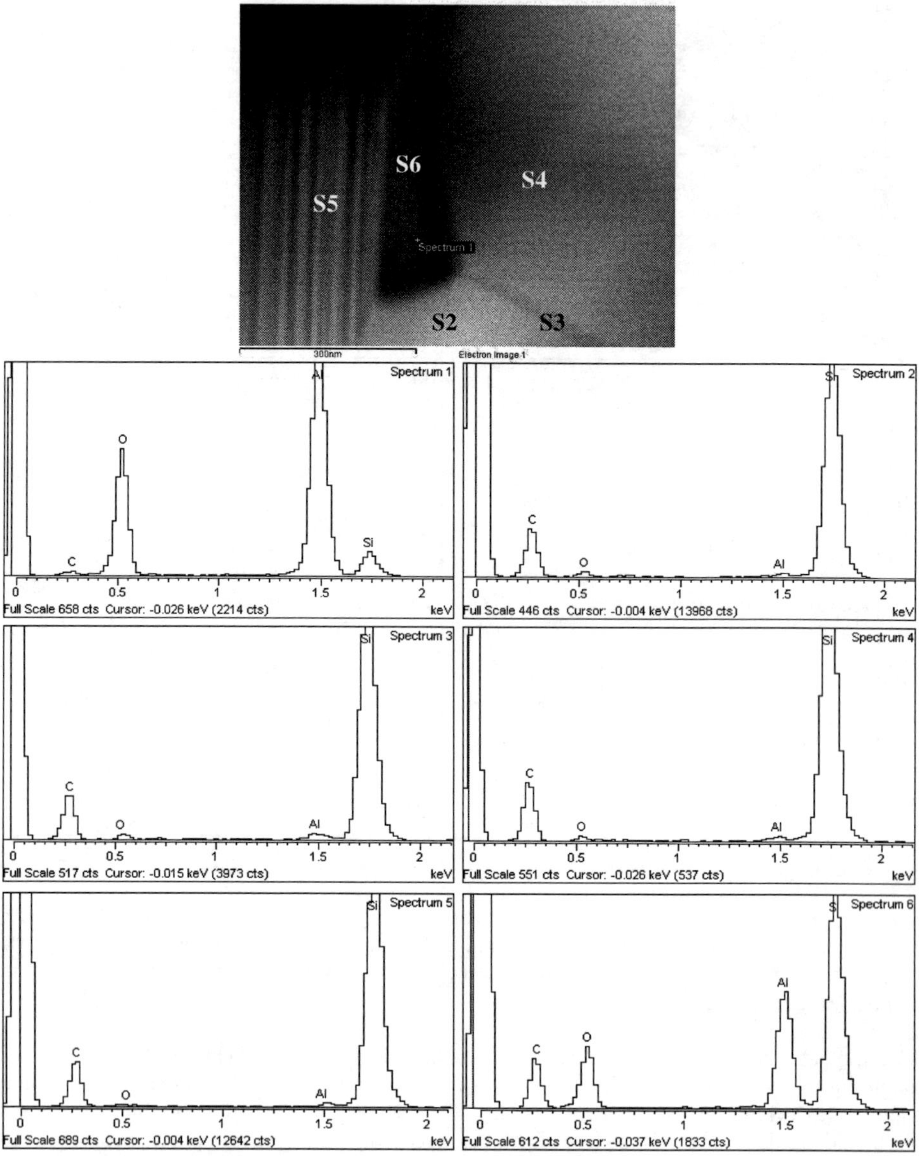

Figure 8. TEM image (top) and six EDS spectra for SiC(Al). Spectrums 1 and 6 are within a triple point, spectrums 2, 4, and 5 are within the grains, and spectrum 3 is at a grain boundary. Al and O are present in triple point and it appears that O is associated with Al along the grain boundaries. Marker is 0.3 μm.

Table III. Summary of the TEM and EDS Characterization of the SiC Materials.

Sample	Triple Junctions	Grain Boundaries	Secondary Phases	Fracture Mode
SiC-N	Al, O	Al, O	None	Intergranular
SiC(Al)	Al, O	Al, O	None	Inter.>Trans.
SiC(Al-B-C)	Al, O	Al	C, B-C rich grains	Transgranular
SiC(AlN)	Al, O	Al, O	None	Intergranular
SiC(AlN-B)	Al, O	Al	C, B-N-rich grains	Trans.>Inter.
SiC(AlN-B-C)	Al, O	Al	C, C-N rich inclusion	Transgranular
SiC(B-C)	None	None	C	Transgranular

SUMMARY

Based on preliminary TEM work it appears that the fracture mode is controlled by the grain boundary chemistry more than the triple point chemistry. All Al or AlN additives contained O and N independent of whether B_4C and C were added due to their ability to reach a state of closed porosity earlier than when no aluminum was present. Materials that fractured intergranularly have shown a wide range of grain sizes, but had both Al and O in the grain boundaries. Samples which fractured transgranularly typically had Al in the grain boundary but little O. This suggests that the wetting characteristics of high-temperature liquid phases likely control the grain boundary chemistry for aluminum additives. More transmission electron microscopy is needed in order to understand the role of grain boundary chemistry in these SiC materials.

ACKNOWLEDGMENTS

This SBIR work was performed for the U.S. Army under contract DAAD17-02-C-0052. Appreciation is expressed to Angela Anderson and Lyle Miller of Ceramatec for help with sample preparation and x-ray diffraction work. Technical discussions with Dr. Svante Prochazka (retired GE), Dr. William Rafaniello, and Dr. Jim Campbell (ARL) are gratefully acknowledged.

REFERENCES

[1] S. Prochazka, "Role of Boron and Carbon in the Sintering of Silicon Carbide," *Proc. of the Conference on Ceramics for High Performance Applications*, Hyannis, Mass., 1973, ed. by J. J. Burke, A. E. Gorum, and R. M. Katz (Brook Hill Publ. Co., 1975).

[2] I. B. Cutler, P. D. Miller, W. Rafaniello, H. K. Park, D. P. Thompson and K. H. Jack, "New Materials in the Si-C-Al-O-N and Related Systems," *Nature*, **275** 434-35 (1978).

[3] W. Rafaniello, M. R. Plichta and A. V. Virkar, "Investigation of Phase Stability in the System Silicon Carbide-Aluminum Nitride," *J. Am. Ceram. Soc.*, **66**[4] 272-6 (1983).

[4] A. Ezis, "Monolithic, Fully Dense Silicon Carbide Material, Method of Manufacturing, and End Uses," U. S. Patent 5,372,978 (Dec. 13, 1994).

[5] R. A. Alliegro, L. B. Coffin, and J. R. Tinklepaugh, "Pressure-Sintered Silicon Carbide," *J. Am. Ceram. Soc.*, **39**[11] 386-89 (1956).

[6] W. Böcker, H. Landfermann, and H. Hausner, "Sintering of Alpha Silicon Carbide with Additions of Aluminum," *Pow. Met.*, **11** 83-85 (1979).

[7] K. A. Schwetz and A. Lipp, "The Effect of Boron and Aluminum Sintering Additives on the Properties of Dense Sintered Alpha Silicon Carbide," *Science of Ceramics,* **10** 149-158 (1980).

[8] J. J. Cao, W. J. Moberlychan, L. C. De Jonghe, C. J. Gilbert and R. O. Ritchie, "*In-situ* Toughened Silicon Carbide with Al-B-C Additives," *J. Am. Ceram. Soc.,* **79** [2] 461-69 (1996).

[9] M. Flinders, D. Ray, A. Anderson, and R. A. Cutler, "High-Toughness Silicon Carbide as Armor," to appear in *J. Am. Ceram. Soc.*, **88** (2005).
[10] R. F. Davis, J. E. Lane, C. H. Carter, Jr., J. Bentley, W. H. Wadlin, D. P. Griffis, R. W. Linton, and K. L. More, "Microanalytical and Microstructural Analysis of Boron and Aluminum Regions in Sintered Alpha Silicon Carbide," *Scanning Electron Microscopy*, **3** 1161-7 (1983).
[11] Y. Tajima and W. D. Kingery, "Grain Boundary Segregation in Aluminum-Doped Silicon Carbide," *J. Mater. Sci.*, **17**[8] 2289-97 (1982).
[12] R. Hanninger, G. Grathwohl, and F. Thuemmler, "Microanalytical Investigation of Sintered Silicon Carbide. Part 1. Bulk Material and Inclusions," *J. Mater. Sci.*, **18**[2] 353-64 (1983).
[13] K. Kaneko, S. Honda, T. Nagano, and T. Saitoh, "Analytical Investigation of Grain Boundaries of Compressive Deformed Al-Doped Sintered β-SiC," *Mater. Sci. & Eng.*, **A285** [1-2] 136-143 (2000).
[14] Z. F. Zhang, M. E. Sixta, and L. C. De Jonghe, "Grain Boundary Evolution in Hot-Pressed ABC-SiC," *J. Am. Ceram. Soc.*, **83**[11] 2813-20 (2000).
[15] X. F. Zhang and L. C. De Jonghe, "Aluminum-Containing Intergranular Phases in Hot-Pressed Silicon Carbide," *J. Mater. Res.*, **19**[8] 2510-16 (2004).
[16] R. Ruh and A. Zangvil, "Composition and Properties of Hot-Pressed Silicon Carbide-Aluminum Nitride Solid Solutions," *J. Am. Ceram. Soc.*, **65**[5] 260-5 (1982).
[17] G. E. Hilmas and T. Y. Tien, "Effect of AlN and Al_2O_3 Additions on the Phase Relationships and Morphology of SiC. Part II. Microstructural Observations," *J. Mater. Sci.*, **34**[22] 5613-21 (1999).
[18] T. Kinoshita, S. Munekawa, and S. I. Tanaka, "Effect of Grain Boundary Segregation of High-Temperature Strength of Hot-Pressed Silicon Carbide," *Acta. Materialia* **45**[2] 801-9 (1997).
[19] H. J. Choi and Y. W. Kim, "Intergranular Film Thickness of Self-Reinforced Silicon Carbide Ceramics," *J. Europ. Ceram. Soc.*, **24**[15-16] 3795-3800 (2004).
[20] R. M. Flinders, D. Ray, A. Anderson, and R. A. Cutler, "Microstructural Engineering of the Si-C-Al-O-N System," this proceedings.
[21] S. Shinozaki, R. M. Williams, B. N. Juterbock, W. T. Donlon, J. Hangas, and C. R. Peters, "Microstructural Developments in Pressureless Sintered β-Silicon Carbide Materials with Aluminum, Boron, and Carbon Additions," *Am. Ceram. Soc. Bull.*, **64**[10] 1389-93 (1985).
[22] A. Zangvil and R. Ruh, "Phase Relationships in the Silicon Carbide-Aluminum Nitride System," *J. Am. Ceram. Soc.*, **71**[10] 884-90 (1988).
[23] R. M. Williams, B. N. Juterbock, S. S. Shinozaki, C. R. Peters, and T. J. Whalen, "Effects of Sintering Temperatures on the Physical and Crystallographic Properties of β-Silicon Carbide," *Am. Ceram. Soc. Bull.*, **64**[10] 1385-9 (1985).
[24] D. Ray, M. Flinders, A. Anderson, and R. A. Cutler, "Hardness/Toughness Relationship for SiC Armor," *Ceram. Sci. and Eng. Proc.*, **24** 401-10 (2003).
[25] M. Flinders, D. Ray and R. A. Cutler, "Toughness-Hardness Trade-Off in Advanced SiC Armor," *Ceram. Trans.*, **151** 37-48 (2003).
[26] R. A. Cutler and T. B. Jackson, 'Liquid Phase Sintered Silicon Carbide," pp. 309-318A in *Third International Symposium on Ceramic Materials and Components for Engines*," ed. by V. J. Tennery (Am. Ceram. Soc., Westerville, OH. 1989).

DEFECT ENGINEERING OF SAMPLES FOR NON-DESTRUCTIVE EVALUATION (NDE) ULTRASOUND TESTING

Raymond Brennan
Rutgers University
607 Taylor Road
Piscataway, NJ 08854-8065

Richard Haber
Rutgers University
607 Taylor Road
Piscataway, NJ 08854-8065

Dale Niesz
Rutgers University
607 Taylor Road
Piscataway, NJ 08854-8065

James McCauley
US Army Research Laboratory
Aberdeen Proving Ground, MD 21005-5066

ABSTRACT

Defect engineered samples were successfully fabricated by introducing hollow polymer spheres into the bulk of pressed Al_2O_3 samples. The polymer spheres burned out at ~525°C during sintering of the samples at 1600°C for three hours. An ultrasound C-scan system was used to nondestructively image the spherical pores within the matrix by gating several regions of the bulk and detecting large attenuation changes in the sample. The C-scans revealed different spherical pore distributions for three different gated thickness regions within the bulk. Microscopy was used to evaluate the cross-section of one of the defect engineered samples, and spherical pores were located, confirming the presence of the engineered defects and adding legitimacy to the results of the ultrasound C-scan study.

MOTIVATION FOR DEFECT ENGINEERING

The feasibility of utilizing high frequency ultrasound non-destructive evaluation (NDE) for detection of micron-size defects and microstructural changes in ceramic armor materials has recently been studied[1,2]. Analysis of critical defect locations, sizes, and distributions in ceramic armor is currently being investigated. Ultrasound C-scans have been conducted on various commercial armor samples and the image data has been quantitatively analyzed in order to differentiate high density armor grade samples from samples containing defects and low density regions[1,2]. While the data received from evaluating commercial samples is valuable, the positions and sizes of the features in these samples cannot be pre-determined. In order to test the integrity of ultrasound as a non-destructive characterization tool, samples must be fabricated with defects of known position, size, shape, and distribution within the bulk matrix. The ability of the system to accurately detect these known features can be used to measure of the limitations of the technique. For this reason, the replication of micron-size features representative of the size and shape of common defects found in ceramic armor samples has been studied. Implementing these features into the bulk of armor materials has also been tested in order to provide standards for assessing the performance of ultrasound NDE.

POLYMER SPHERE CHARACTERIZATION

Two different sizes of hollow polymer spheres were obtained from Akzo Nobel[3] in order to fabricate bulk defects of various sizes in known locations. The first set of spheres (Sphere Type A) was reported to have a particle size between 20 and 40 microns and the second set of spheres (Sphere Type B) was reported to have a particle size between 60 and 90 microns. The spheres were characterized by scanning electron microscopy (SEM) and thermal gravimetric analysis (TGA) in order to confirm sphere size distributions and to determine the temperature at which the spheres would burn out of the matrix. Figure 1 shows SEM micrographs from both Sphere Type A and Sphere Type B. Both micrographs demonstrate the spherical morphology of the hollow polymer spheres. The micrograph of Sphere Type A shows a majority of spheres within the 15 to 25 micron size range, with several spheres as small as 5 microns and several spheres as large as 35 microns. The micrograph of Sphere Type B shows a majority of spheres within the 60 to 75 micron size range, with several spheres as small as 50 microns and several as large as 90 microns. TGA data was collected up to 700°C for Sphere Type A and Sphere Type B to study the temperature at which there was no longer any weight fraction percent of spheres remaining (Figure 2). For Sphere Type A, the weight fraction percent reached a minimum at ~475°C. For Sphere Type B, the weight fraction percent reached a minimum at ~525°C. This study showed that the polymer spheres would burn out at these respective temperatures.

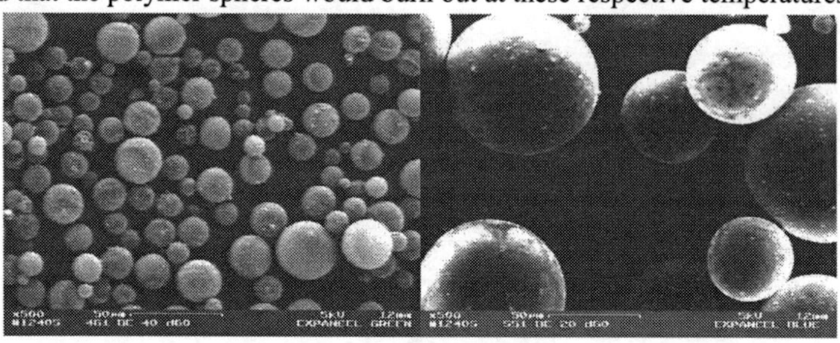

Figure 1. SEM micrographs of Sphere Type A (left) and Sphere Type B (right).

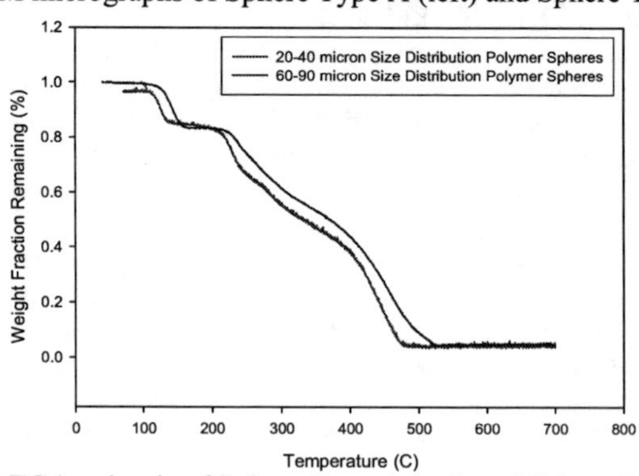

Figure 2. TGA anlaysis of Sphere Type A (red) and Sphere Type B (blue).

DEFECT ENGINEERED SAMPLE PROCESSING

To test the ability of the ultrasound C-scan system to detect specific defects placed within a bulk armor material sample, Al_2O_3 samples containing the hollow polymer spheres were fabricated. Alcoa A16SG[4] alumina powder, which is a commonly used in ceramic armor, was added to a one-inch diameter dry pressing die. Half of the powder was added to the die, and a small amount of polymer spheres was dispersed over the powder bed. Since the polymer spheres were in the micron-size range, they had to be added and controlled very carefully so they could be properly dispersed within the center of the powder bed. The remaining amount of A16SG powder was then added to fill the die, and the sample was dry pressed at a pressure of 100 MPa. Samples were made with both Sphere Type A and Sphere Type B embedded in the bulk. The green samples were sintered at 1600°C for three hours in order to reach full densification. The samples were then polished with a 125-grit diamond wheel to provide flat, parallel surfaces before ultrasound evaluation.

ULTRASOUND C-SCANS OF DEFECT ENGINEERED SAMPLES

Ultrasound C-scan imaging is a technique that uses acoustic waves to discern material and microstructural changes in a bulk sample. In this type of system, a pulser-receiver unit sends an initial electrical pulse to an ultrasonic transducer, and the piezoelectric crystal in the transducer converts the pulse into sound waves. These acoustic waves have an initial frequency that is dictated by the size of the piezoelectric crystal. The waves are sent into the material that is being analyzed, and either transmit, reflect, or scatter energy, depending on the acoustic impedance differences between material boundaries. For a larger difference in acoustic impedance, more energy is reflected or scattered and less energy is transmitted into the material. The scattered and reflected acoustic waves from the material are received by the same transducer, and the piezoelectric crystal converts the signals back into electrical pulses, which are sent to the receiver of the pulser-receiver to amplify the signals. The amplified signals are displayed as a voltage verse time trace on an oscilloscope, which is used to identify the trigger signal from the transducer, the signal from the top surface of the sample, the attenuated signal from the bottom surface of the sample, and additional reflections and echoes. Any detectable defects or features within the bulk of the sample will occur between the top and bottom surface reflections.

In a C-scan, the transducer is rastered over the desired sample area, and the collected signals are assigned to x and y coordinates. The changes in peak amplitude of the signals are evaluated over the selected area. The step size of the scan dictates the smallest increment over which a data point is collected. The gated region or regions dictate which signals are analyzed as the scan is run. Some examples of gated regions include the top surface peak reflection, which is gated to study the amplitude differences at the surface of the sample, the bottom surface peak reflection, which is gated to study density variations and look for defects through the entire bulk of the sample, and the bulk gated region, which is set within a certain thickness of the sample to identify peaks from defects within the bulk. The collected pulse amplitude data is assigned to a color scale, which represents changes in the amplitude over the gated region. Depending on the assigned scale, one color shade difference represents a difference in voltage, commonly several mV per shade. The data is mapped in relation to both the assigned color scale and the x and y positions of the points to produce an image map of the sample over the gated signal.

C-scan imaging was performed at Rutgers University using an ultrasound system manufactured by The Ultran Group[5]. The defect engineered Al_2O_3 sample was scanned with a

10 MHz longitudinal immersion transducer. The sample was placed into an immersion tank and the z-position was manually adjusted until the maximum signal was obtained. The top and bottom surface peak reflections were identified, and a total of five different scans were run over five separate gated regions. First, the top surface signal was gated so that the variation of signal amplitude could be collected over the surface of the sample. The maximum amplitude of the signal was found to be approximately 140 mV in the center of the sample, and the position was changed randomly to get an idea of how the pulse amplitude would vary across the sample. It was determined that an amplitude range of 0 to 150 mV would be sufficient, so the color or gray scale was set to reflect these values. In this case, an amplitude of 0 mV represented and dark blue color (lowest possible amplitude or no signal) and an amplitude of 150 mV represented a dark red color (highest possible amplitude or strong signal). Any color shades in between represented a scale of intermediate amplitudes, which could be used to determine attenuation differences in the sample. For the top surface signal scan, the amplitude distribution was very uniform, with amplitude changes over a very small range (Figure 3). This was the expected result for the polished top surface of the sample. Next, the bottom surface signal was gated, which showed amplitude changes through the entire bulk of the sample. This scan revealed some lighter colored lines extending from the perimeter of the sample which are believed to be radial cracks. There were also some lighter colored circular features of lower amplitude (light blue and yellow) that are believed to show evidence of the spherical pores from the burn out of polymer spheres (Figure 3).

The next three gated regions were from the bulk, the first from the upper region of the bulk, the second from the direct center of the bulk, and the third from the lower region of the bulk. For these scans, a change in acoustic impedance between the Al_2O_3 bulk material and a spherical pore, for example, would result in a peak and a change in color representative of the degree of amplitude change. In this case, a greater difference in acoustic impedance from the presence of a pore would represent a higher intensity amplitude peak and, therefore, a darker color (where red is darkest color on the scale). Figure 3 shows the three gated bulk C-scan images from an alumina sample containing Type B Spheres (60-90 μm). The upper bulk scan shows the distribution of pores (dark red areas) within that thickness of the sample. The middle bulk scan shows a higher degree of spherical pores in a different distribution than the upper bulk region. The lower bulk scan shows yet another distribution of pores throughout the thickness gated toward the bottom of the sample. The dark red features are believed to be the engineered spherical pores from the burn out of polymer spheres at around 525°C because they show the greatest amplitude difference (greatest change in acoustic impedance) from the Al_2O_3 matrix and they are characteristic of the expected pore sizes, shapes, and locations within the bulk. To confirm this, a cross-section of the sample was cut to locate any spherical pores within the dense alumina matrix, and the region was examined using SEM. Figure 4 shows a micrograph displaying one of the spherical pores that was found in the sample, confirming the relationship between the features found by ultrasound C-scan and those identified by SEM. Further experimentation will be conducted to place the polymer spheres in specific patterns within the microstructure, and the optimized technique will also be applied to hot-pressed silicon carbide samples to demonstrate defect engineering in another ceramic armor material.

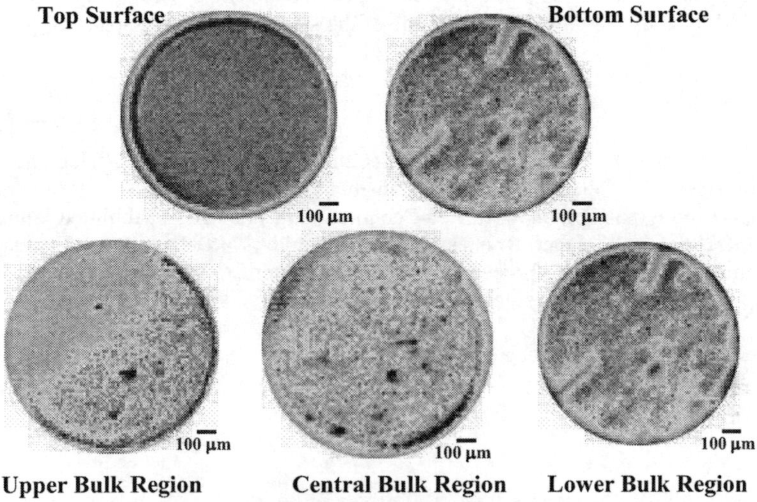

Figure 3. Ultrasound C-scan images of top surface, bottom surface, upper bulk region, central bulk region, and lower bulk region.

Figure 4. SEM micrograph of spherical pore from cross-section of defect engineered sample.

CONCLUSIONS

Defect engineered samples were successfully fabricated by incorporating hollow polymer spheres into a matrix of Al_2O_3. Upon sintering to high temperature, the polymer spheres burned out to leave behind spherical pores within the bulk of the Al_2O_3 samples. Ultrasound C-scans were collected for the samples over five different gated regions – top surface, upper bulk region, central bulk region, lower bulk region, and bottom surface. The C-scan images revealed various spherical pore distributions within the three gated bulk regions. The size, shape, and location of the features, which had amplitudes that were drastically different from the Al_2O_3 matrix led to the conclusion that these features were the engineered spherical pores. This was confirmed by using SEM to study the cross-section of one of the samples, revealing spherical pores within the bulk.

REFERENCES

[1]R. Brennan, R. Haber, D. Niesz, J. McCauley, and M. Bhardwaj; "Non-Destructive Evaluation (NDE) of Ceramic Armor: Fundamentals"; Proceedings of The 29th International Conference and Exposition on Advanced Ceramics and Composites; submitted January 2005.

[2]R. Brennan, R. Haber, D. Niesz, and J. McCauley; "Non-Destructive Evaluation (NDE) of Ceramic Armor: Testing"; Proceedings of The 29th International Conference and Exposition on Advanced Ceramics and Composites; submitted January 2005.

[3]Akzo Nobel Expancel Microspheres (http://www.expancel.com)

[4]Alcoa (http://www.alcoa.com)

[5]The Ultran Group (http://www.ultrangroup.com)

DEFINING MICROSTRUCTURAL TOLERANCE LIMITS OF DEFECTS FOR SiC ARMOR

Memduh Volkan Demirbas
Rutgers University
Ceramic and Materials Engineering
607 Taylor Road
Piscataway, NJ 08854

Richard A. Haber
Rutgers University
Ceramic and Materials Engineering
607 Taylor Road
Piscataway, NJ 08854

ABSTRACT

Near theoretical density, >99.5%, has generally been used to assess structural integrity and is commonly used as a key acceptance criteria for armor product selection. However, it is unknown whether a slight change in, residual porosity, the shape and dispersity of pores are detrimental to the dynamic performance of the product. This study aims to define the upper limit of acceptable defects (i.e. porosity and inclusions) and to quantify the microstructure of SiC armor. The key goal within this study is to develop an understanding of the factors critical to the design, fabrication and testing of micro/macro-tailored ceramics with enhanced multi-hit ballistic performance. Commercial SiC samples with varied densities (3.0-3.2 g/cm^3) have been obtained and microstructural assessment on polished sections has been performed. Pore size distribution and average pore size have been calculated using image analysis. Nearest neighbor distance distributions of different microstructures have been found. Tessellation analysis has been performed to find out the degree of inhomogeneity in microstructures. Finally, pair correlation functions have been determined.

INTRODUCTION

The origin of microstructural anomalies and the effect of these features on the properties of bulk ceramics are of great importance to armor materials. The properties of a material are determined by the kinds of components present and their geometry. Thus, geometrical analysis can provide valuable tools to understand the formation of microstructure and final material properties. At this point, stereology comes as an important method to provide geometrical analysis on microstructures. The following study has been aimed at performing geometrical analysis to determine the homogeneity of pores in SiC.

Different types of defects, such as carbon inclusions, metallic inclusions, pores could be found in sintered or hot-pressed materials. The emphasis in this paper is given to pores among the others. Pore location is an important concept as well as average pore size and pore size distribution. Spatial and planar distribution is often times overlooked however the variations in density in a sample might be detrimental to the properties such as elastic modulus, hardness and strength of a material. Microstructural investigation is a good way to assess these variations in density, which is a result of pores, their location and possible clustering.

Spatial distribution of pores in any material could be shown in many ways. A material might have evenly distributed and relatively small pores or larger pores as can be seen in Figure 1(a) and (b), respectively. Large and elongated pores as in Figure 1(c) might be present which will certainly affect mechanical properties. Isolated large pores (Figure 1(d)) or even clusters of pores (Figure 1(e)) are other possible patterns among many different ones that have not been mentioned here. To show the differences by quantification of microstructures and correlating these with mechanical properties will be important for evaluation of armor materials in the future.

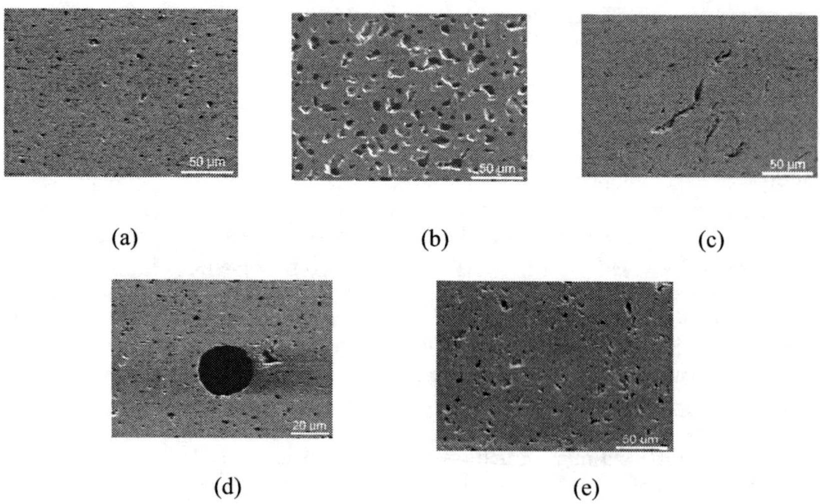

Figure 1. (a) relatively small pores (b) larger pores (c) elongated pores (d) isolated large pore (e) clusters of pores

PORE SIZE DISTRIBUTION

Silicon carbide samples were used in this study. They were cut using a diamond blade and water as a coolant. They were mounted in epoxy, ground using 125, 40, 20, 10 µm diamond grits and polished with 6 and 1 µm diamond suspensions. The images representative of each sample is given in Figure 2.

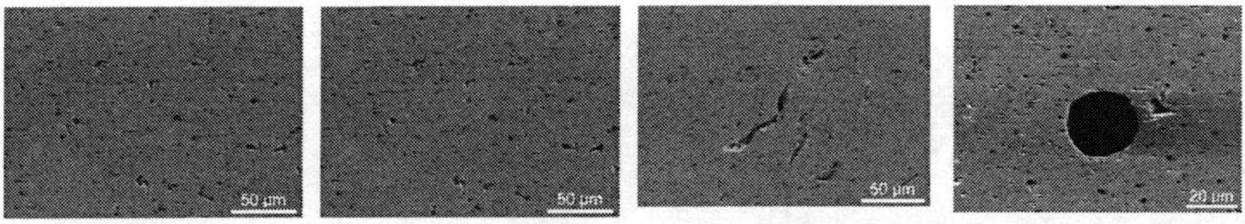

Figure 2. (a) Sample A (b) Sample B (C) Sample C (d) Sample D

Ten pictures are taken using secondary electrons by LEO-Zeiss Gemini 982 Field Emission Scanning Electron Microscope, 5 of them at 500× magnification and 5 at 1000×. The images are taken in a way to represent the microstructure of the whole sample. Polished sections are used for this part of the study in order distinguish pores easily.

Image analysis has been used to obtain quantitative information on the microstructures. Image Processing Toolkit 3.0 has been used for image analysis.

First step of the image analysis is thresholding, which is when a pixel's value is greater than a certain number; its value is substituted by 1. If its value is less than or equal to that number, it is replaced by zero. A binary image is obtained after thresholding step. This step is followed by closing operation, which is defined as a dilation operation followed by an erosion operation. The purpose of this operation is to smooth features and to remove isolated pixel noise from the image. The next step after closing is filling holes. This is based on the definition that a hole is defined as an isolated region in a feature with background color. This operation adds inner parts of the pores, which might be excluded due to thresholding. The final step is cutting off edge-touching features in order to include only the features in the field of view[1, 2]. Finally, a report on the features is obtained, which provides useful information such as area, equivalent diameter, form factor, x- and y- coordinates of each feature in the image for further analysis.

Average pore size and pore size distribution of four samples are calculated using image analysis. Equivalent diameter of features is used for calculation of average pore diameter. Pore size distribution is given in Figure 3 and average pore size is given in Figure 4. "More" part in the x-axis of Figure 3 refers to pores larger than 10 μm. Among all samples, B has the smallest pore size with an average value of 1.01 μm. Samples C and D have an average pore size of 1.05 μm and A has the largest average pore size with 1.28 μm. However A has a much lower standard deviation, ±0.89 μm, when compared to C and D, ±1.43 and ±1.57, respectively. These values show that samples C and D have a broader pore size distribution than samples A and B, which can be seen in Figure 4.

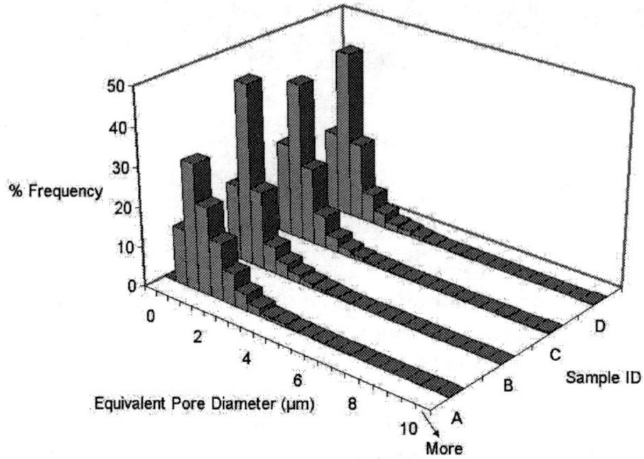

Figure 3. Pore size distribution by stereology for the four samples

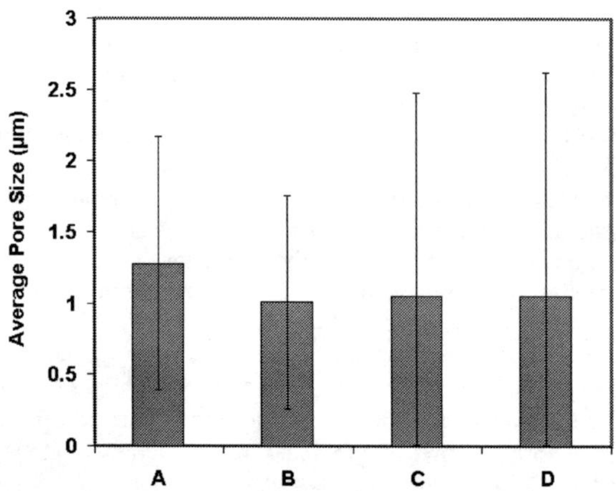

Figure 4. Average pore size for the four samples

THE NEAREST-NEIGHBOR DISTANCE DISTRIBUTION

Average pore size and pore size distribution does not give sufficient information about the microstructure of a material. Additional information is needed to comment on a microstructure. The focus in this study is on the location of pores and their spatial distribution, therefore more information should be extracted from the images.

One method to assess spatial distribution of pores is to define nearest neighbor distance distribution and compare the observed and mean and variance for a random Poisson distribution[3].

If P(R) is defined as density function, then the nearest neighbor distribution function of pores in a plane is described so that P(R)·dR is equal to the probability that there is no other pore centroid in R distance and there is at least one pore centroid in (R+dR) distance. Then the average nearest neighbor distance is calculated by[4];

$$\langle R \rangle = \int_0^\infty RP(R)dR$$

If the points are randomly distributed in a plane, then ⟨R⟩ is equal to $\{0.5/\sqrt{N_A}\}$, where N_A is the number of points per unit area. Second, third, n-th nearest-neighbor distributions can also be calculated by using a similar approach[4, 5, 6].

For estimation of the nearest-neighbor distance distribution, (x,y) centroid coordinates of the pores observed on the ceramographic plane are measured by Image Processing Toolkit 3.0. Then, the distance between (x_1,y_1) and all the other pores $[(x_2,y_2), (x_3,y_3)\ldots]$ seen in the microstructure is measured and the shortest distance is recorded. This procedure is repeated for all the pores and then the nearest-neighbor distance distribution is obtained. The nearest-neighbor distance distribution for SiC samples is given in Figure 5.

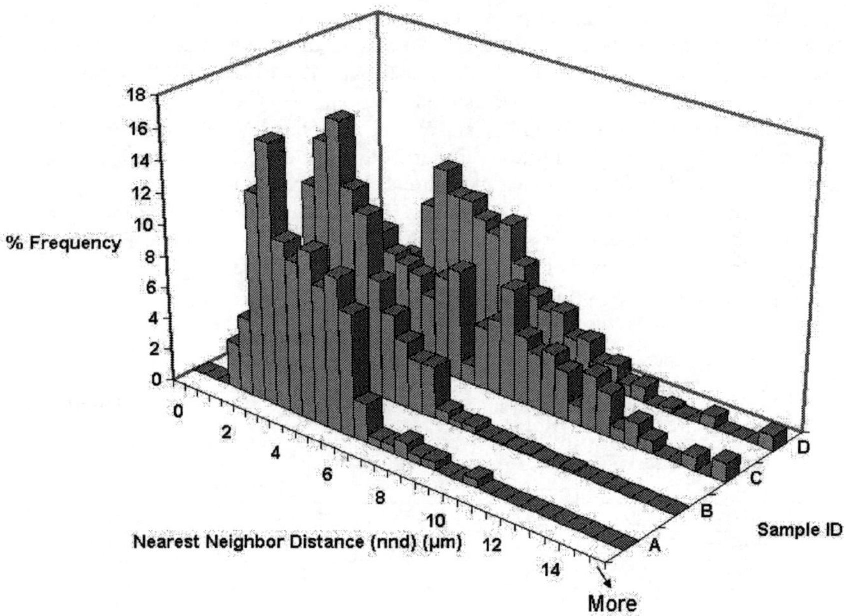

Figure 5. Nearest neighbor distance distributions

Sample A and B show relatively narrow nearest neighbor distance distributions with small values for variance of the distributions, 2.32 and 2.09 μm^2, respectively. This narrow distribution of nearest neighbor distances *(nnd)* can be observed by looking at the right corner of the histogram in which the % frequency of distances longer than arbitrary 15 μm is given and it is zero for both sample A and B. 'More' refers to values larger than 15 μm for the *nnd* distributions. However, samples C and D show different behavior. Broader *nnd* distribution can be seen easily for both samples, with variance values 10.83 and 6.97 μm^2, respectively. The other important observation for these two histograms is that there is considerable amount of *nnd* values on the right side of the histogram. This is a clear indication of inhomogeneous microstructures that these two samples possess.

To be more quantitative, in terms of centroid distributions, it has been suggested[5] that the nature of a distribution may be described by the parameters Q and V, where;

$$Q = \frac{\mu_o}{\mu_e} = \frac{\text{observed mean nearest-neighbor distances}}{\text{expected mean nearest-neighbor distances}}$$

$$V = \frac{\text{var}_o}{\text{var}_e} = \frac{\text{observed variance of nearest-neighbor distances}}{\text{expected variance of nearest-neighbor distances}}$$

The expected mean and variance are for a random distribution of events. Different types of distributions can be determined by these two parameters and the conditions for these distributions are as follows: (a) random distributions, $Q \approx 1$ and $V \approx 1$; (b) regular distributions, $Q > 1$ and $V < 1$; (c) clustered distribution, $Q < 1$ and $V < 1$; and (d) random distribution with clusters, $Q < 1$ and $V > 1$[5, 7, 8]. Q and V values for each sample are determined and given in Table I. Observed mean and variance nnd values are calculated from the distributions given in Figure 4. Expected values are calculated by the Image Processing Toolkit 3.0. The Q values for all four samples are approximately equal to 1; however the V values differ strongly from the random case. Sample B is the only one whose V value is approximately 1. Other three samples, especially C and D, have V values way off the idealized value, ~1.

Table I. Q and V table

	Mean nnd (μm)	Variance nnd (μm)	Mean nnd (Random) (μm)	Variance nnd (Random) (μm)	Q	V	Distribution Type
A	7.550	9.029	6.542	2.423	1.154	3.726	Random
B	3.769	2.103	3.262	1.208	1.155	1.174	Random
C	6.093	10.798	5.753	2.131	1.059	5.068	Random
D	5.167	6.971	4.674	1.731	1.104	4.027	Random

TESSELLATION ANALYSIS

This method is based on constructing polygons around features on microstructures. Tessellation analysis provides information on the location of features relative to each other so it is a suitable technique to detect clustering in a microstructure. The polygons are constructed so that a line is drawn at midway between two features. As this is done for all pores, tessellated microstructure is obtained at the end as it can be seen in Figure 6.

Tessellations given in Figure 6 give cell area distributions as shown in Figure 7, which provide information on the homogeneity of the microstructure. This is a good way to assess microstructures since the cell area is dependent on the whole surrounding environment of the pore concerned. As explained previously, 'more' in Figure 7 refers to cell areas larger than 100 μm^2 .The distribution for sample A is very broad in between the limits 0-100 μm^2 but the cell area percentage above 100 μm^2 is low when compared to samples C and D. Sample B has the narrowest distribution among all and the cell area percentage larger than 100 μm^2 is minimal.

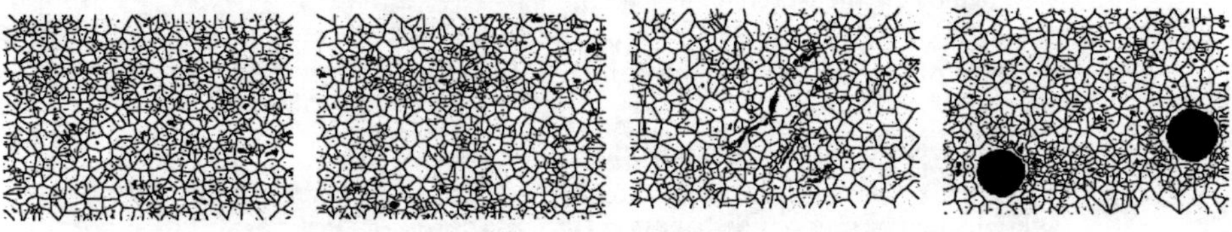

(a) (b) (c) (d)

Figure 6. Tesellations for (a) Sample A, (b) Sample B, (c) Sample C, (d) Sample D

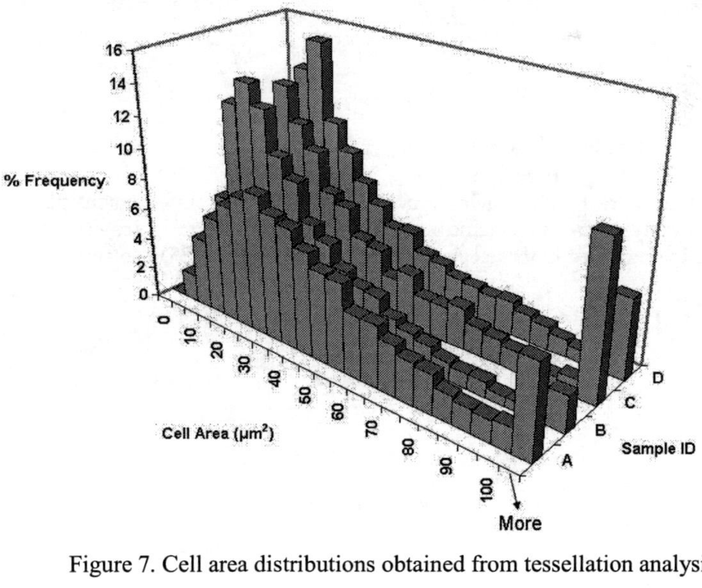

Figure 7. Cell area distributions obtained from tessellation analysis

Other than these distributions, two parameters, P_1 and P_2, have been identified by Murphy *et al.*, which help to quantify the microstructure in terms of clustering. P_1 is based on the variance of the cell area distribution. It is a parameter that the extent of spread in a distribution can be found since the variance from a real microstructure is compared to that of a random microstructure. P_2 is a measure of skewness, which shows the asymmetry in a distribution. If there is a tail in the distribution, that would increase the skewness of that distribution. The extraction of parameters P_1 and P_2 is given in following equations[9].

$$P_1 = \frac{V}{V_{random}}$$

where;

variance of the measured areas (μm^4), $V = \frac{1}{n}\sum_{i=1}^{n}\left(A_i - \overline{A}\right)^2$

variance of random distribution (μm^4), V_{random}

average measured cell area (μm^2), $\overline{A}$

average area of random distribution (μm^2), $\overline{A}_{random}$

$$P_2 = \frac{S}{S_{random}}$$

where;

skewness of measured areas, $S = \dfrac{\frac{1}{n}\sum_{i=1}^{n}\left(A_i - \overline{A}\right)^3}{V^{3/2}}$

skewness of random distribution, S_{random}

The parameters P_1 and P_2 have been calculated for reinforcement particles in MMCs by Murphy *et al.*, and same principles are applied for pores in SiC armor. Random distributions used in these calculations are simulated by STG software.

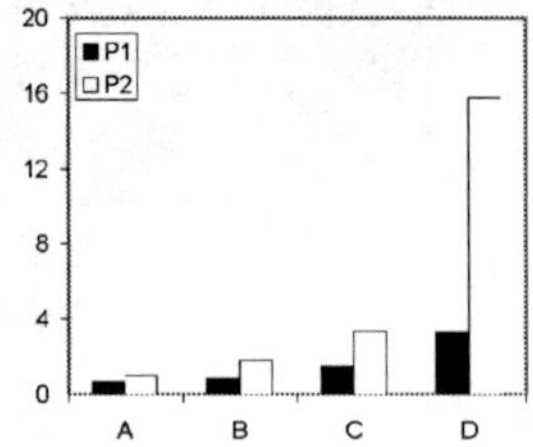

Figure 8. P_1 and P_2 parameters for SiC sample

P_1 and P_2 parameters are also calculated for all samples, which are given in Figure 8. Sample D has the largest values, 3.28 and 15.78, respectively, as opposed to 1, which is the expected value for random distribution of particles. These two parameters are supposed to quantify clustering[9]; however one clarification is required here. The particles concerned in that study has uniform size however there are clearly larger pores in our samples, so the contribution of large pores leading to large cell areas should not be misleading. However, it is fair to say that distribution of pores throughout sample D is not uniform and there is some degree of clustering.

PAIR CORRELATION FUNCTIONS

Another way to characterize microstructures and to learn about the spatial distribution of features is to use *pair correlation functions*. When a planar case is considered and if λ denotes the planar point density, the mean number of points per unit area, then the probability to find a point in a small circle of area dA is λdA. If two small circles with areas dA_1 and dA_2 and center-to-center distance, r, are considered, P(r) is the probability that the two circles contain both a point of the point process and it can be written as follows:

$$P(r) = \lambda^2 \cdot g(r) \cdot dA_1 \cdot dA_2$$

g(r) in this equation is called *pair correlation function*[10].

For a completely random point process g(r) = 1 for all r since the events to have points in the two circles are independent[4, 10].

Pair correlation functions regarding the samples are given in Figure 9. The different curves in each graphs represents estimation from different micrographs. Most of plots in each graph are dominated by curves which rise monotonically to its asymptotic value and is distinguished by relative lack of structure. Some of the curves for sample C and D have sharp peaks very close the y-axis. Strong first peak at a distance well below the mean interparticle distance (r_0) is an indication of clustering. The curves for sample A and B are typical random distribution curves.

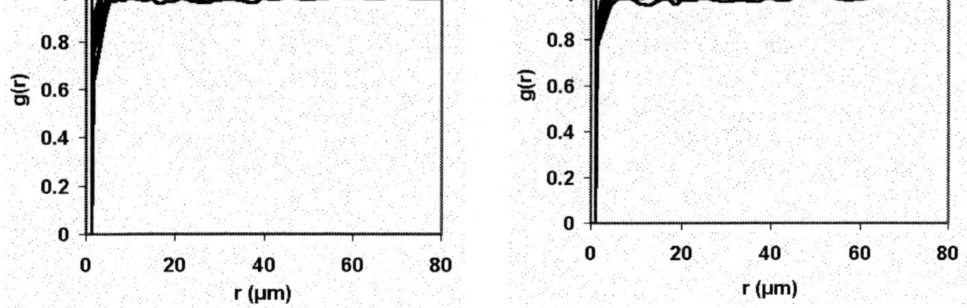

(a)

(b)

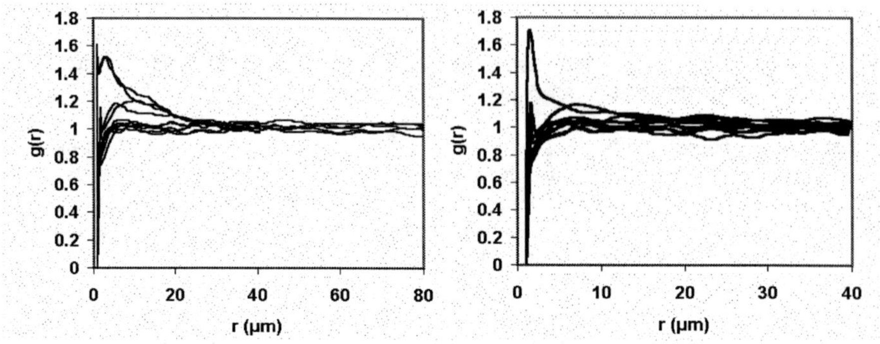

(c) (d)

Figure 9. Pair correlation functions of (a) Sample A, (b) Sample B, (c) Sample C, (d) Sample D

CONCLUSION

Consequently, four different SiC samples are investigated in terms of spatial distribution of pores and a number for inhomogeneity of microstructure is assigned by using three different but closely related approaches. The term "inhomogeneity" is used at this point because of the broad size distribution of pores. Some parameters supposedly showing "degree of clustering" has also contribution from that broad size range. The ultimate purpose of this study is to quantify and assign a number for every microstructure possible. Then mechanical tests will follow and a correlation between the two will be obtained as the continuation of this study.

REFERENCES

[1]John C. Russ, "Computer-Assisted Microscopy" 1990 Plenum Press, New York

[2]M. Berman, L. M. Bischof, E. J. Breen, G. M. Peden, "Image Analysis" *Materials Forum* 18, 1-19 (1994)

[3]George F. Vander Voort, "Computer-Aided Microstructural Analysis of Specialty Steels" *Materials Characterization* 27, 241-260 (1991)

[4]P. Louis, A. M. Gokhale, "Application of image analysis for characterization of spatial arrangements of features in microstructures", *Metallurgical and Materials Transactions A*, Vol 26A, 1449-1456, June 1995

[5]J. P. Anson, J. E. Gruzlezski, "The Quantitative Discrimination between Shrinkage and Gas Microporosity in Cast Aluminum Alloys Using Spatial Data Analysis" *Materials Characterization* 43, 319-335 (1999)

[6]Joachim Ohser, Frank Mucklich, "Statistical Analysis of Microstructures in Materials Science" 2000 John Wiley & Sons, Ltd.

[7]P. A. Karnezis, G. Durrant, B. Cantor, "Characterization of Reinforcement Distribution in Cast Al-Alloy/SiC$_p$ Composites" *Materials Characterization* 40, 97-109 (1998)

[8]N. Yang, J. Boselli, P. J. Gregson, I. Sinclair, "Simulation and Quantitative Assessment of Finite-size Particle Distributions in Metal Matrix Composites" *Materials Science and Technology* 16 (7-8): 797-805 JUL-AUG 2000

[9]A. M. Murphy, S. J. Howard, T. W. Clyne, "Characterization of severity of particle clustering and its effect on fracture of particulate MMCs" *Materials Science and Technology,* 14 (9-10):959-968 (1998)

[10]J. Hubalkova, D. Stoyan, "On a qualitative relationship between degree of inhomogeneity and cold crushing strength of refractory castables", Cement and Concrete Research 33, 747-753 (2003)

Fracture Mechanism of Ceramic Armor

CHARACTERIZATION OF SUBSURFACE DAMAGES OF STATIC AND DYNAMIC INDENTED ARMOR CERAMICS: SiC AND Si_3N_4

Jong Ho Kim, Young Gu Kim, and Do Kyung Kim
Department of Materials Science and Engineering
Korea Advanced Institute of Science and Technology
Yusong, Taejon, 305-701, Korea

ABSTRACT

Static and dynamic indentation damage evolution was investigated in silicon carbide and nitride using sphere indentation and explosive indentation techniques. Different microstructures of SiC and Si_3N_4 were prepared and the subsurface damage of the indented specimen was observed by using the bonded-interface technique. The results of static indentation shows that the microstructural heterogeneity determines the primary fracture mode of SiC and Si_3N_4: brittle fracture is dominant in fine-grained microstructure and quasi-plastic deformation in coarse-grained one. Dynamic indentation, however, generates massive quasi-plastic deformation zone in silicon carbide and silicon nitrides, except solid-state sintered SiC. Electron micrographs reveal that the major damage source of quasi-plastic zone is grain boundary defeat in both case of SiC and Si_3N_4.

INTRODUCTION

Silicon carbide and nitride are most successful non-oxide ceramics due to its high hardness, high strength, low thermal expansion coefficient, and high fracture toughness. These materials have been successfully demonstrated in a variety of applications, such as cutting tools, bearings, gas turbine engine components, and ceramic armor.[1] The effect of microstructure on mechanical properties is crucial importance for performance and reliability of ceramic components.[2] Static indentation on brittle ceramics has been extensively studied because it is simple and powerful means to reveal the effect of microstructure on elastic and plastic deformation of ceramics.[3] It is well-known that fine-grain ceramics show brittle dominant facture behavior and elongated grained ceramics exhibit quasi-plastic deformation. The control of microstructure enables optimization of brittleness and quasi-plasticity of ceramics for each application. In case of silicon carbide and nitride, mechanical properties, especially fracture toughness have been greatly improved by microstructure tailoring.[4]

When silicon carbide and nitride are used as armor components, the dynamic impact loading condition is subjected which usually results a massive catastrophic failure through the localized damage accumulation. Fracture behavior during the dynamic impact can differ significantly from static loading. Several techniques were suggested to investigate the dynamic fracture of ceramics and to explain the deformation of ceramics at high strain rate.[5] Attempts to find the relation between static and dynamic fracture are still in under investigation. In recent preliminary study about observation of static and dynamic indentation damage mode of several ceramics at microstructural level, a similar fracture tendency was found.[6,7,8] This result implies that dynamic indentation damage of ceramics can be explained as a similar failure mode of static indentation. However previously reported study was insufficient to conclude the relation between static and dynamic fracture modes.

In this study, static and dynamic indentation on silicon carbide and nitride were conducted to elucidate the damage process responsible for the static and dynamic failure. Sphere and explosive indentations were adopted to provide the extremely different condition of strain rate in two grades of silicon carbide and three grades of silicon nitrides. Subsurface damage of indented specimens was carefully observed with optical and scanning electron microscope. Primary failure mode of static and dynamic indented silicon carbides and silicon nitrides was determined. In addition, transmission electron microscopy was applied to characterize the micromechanical failure of dynamically damaged specimens. The effect of microstructure of silicon carbides and silicon nitrides on static and dynamic indentation damage have been discussed.

EXPERIMENTAL

Specimen Preparation

Two grades of silicon carbide were used for this study. One was fabricated through hot pressing (HP) and the other by the solid state sintering (SS). The sintering process strongly affects the microstructure and phase compositions. Hot pressed specimen contains primarily all α with a small amount of β phase and a glassy grain boundary phase and solid stated sintered contains only a phase and no liquid grain boundary.

Three grades of silicon nitride was prepared by hot pressing. Densification was achieved in nitrogen gas at 1 atm under uniaxial pressure 25 MPa, at 1600, 1700, and 1800 °C for 1 hour. These sintering temperatures produced fine equiaxed (F), intermediate (M), and coarse elongated (C) microstructure. Detail processing conditions were described earlier.[4]

Specimen surfaces were polished down to 1 μm diamond slurry finish to enable characterization of microstructure. These surface were plasma etched to reveal the grain structure, however SS-SiC is hardly etched by plasma environment so that only fracture surface was used to characterize the microstructure. The microstructure of specimens was observed by scanning electron microscope. Figure 1(a) and (b) show microstructure of SS-SiC and HP-SiC. Silicon nitrides, Fig. 1(c), (d) and (e), have bimodal microstructure with different ratio of α and β: F-Si_3N_4 with ~75 % 0.5 μm α grains and ~ 15 % β grains, M- Si_3N_4 with ~78 % 0.5 μm wide and 4 μm long β grains and ~12 % of α grains, C-Si_3N_4 having 100 % 1.5 μm wide and 9.0 μm long β grains. All silicon nitrides has intergranular glassy grain boundary phase covering all grain boundary. Table 1 shows the basic physical properties of specimens for this study. Young's modulus and Poisson's ratio were determined for each sample using routine impulse excitation technique. The Vickers indentations were used to determine the hardness of materials, using equation $H = P/2a^2$, where P is load and a is half-diagonal. In order to measure the strength, the flexural tests were conducted with four-point bending configuration. Another set of bend tests was made on specimens after Vickers indentation to estimate the toughness, with equation $T_0 = 0.62(E/H)^{1/8}(\sigma P^{1/3})^{3/4}$ where E is modulus, H is hardness, σ is strength, and P is load.

Static and dynamic indentations

Static sphere indentation was applied to induce contact damage. Prior to testing, the specimens were machined into 6 mm x 8 mm x 20 mm rectangular-shaped bars. The top surface of specimens was grinded and polished down to 1 μm diamond slurry finish. Side surfaces normal to the top surface were also polished to observe subsurface damage. The opposing side

Table I Processing and physical properties of silicon carbides and silicon nitrides

Material	SS-SiC	HP-SiC	F- Si_3N_4	M-Si_3N_4	C-Si_3N_4
Processing condition	Solid-state sintering	Hot-press	Hot-press at 1600 °C	Hot-press at 1700 °C	Hot-press at 1800 °C
Hardness, H (GPa)	29.1	19.5	20.6	16.5	15.9
Toughness, T(MPa $m^{1/2}$)	2.46	3.75	3.8	5.3	4.9-7.3 †
Strength, σ_F (MPa)	523	525	885	1084	792
Young's modulus, E (GPa)	440	442	331	326	314
Poisson's ration, ν	0.17	0.17	0.27	0.28	0.29

† R-curve behavior is observed

surfaces were polished and adhered together using epoxy to fabricate 'bonded-interface' specimens.[9] The gaps between two bars were maintained below 10 μm to minimize the additional free surface effect. After indentation, these bonded interface specimens were separated and side surface was cleaned by acetone.

Explosive indentation testing was conducted on same bonded interface specimens to

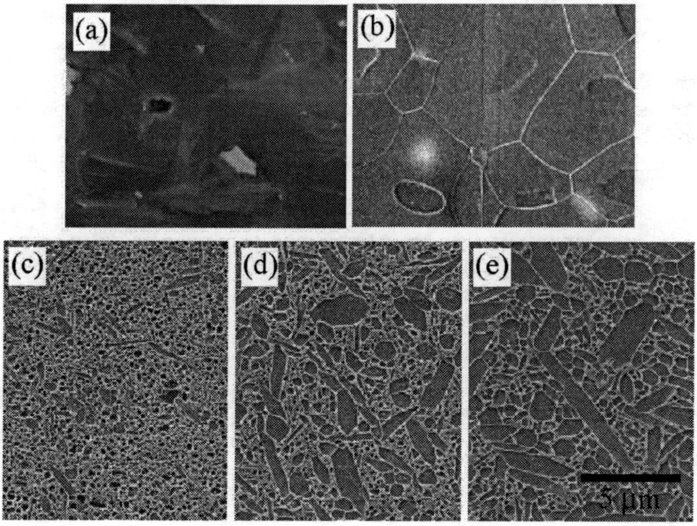

Figure 1 SEM micrographs of polished and plasma-etched silicon carbide and nitride specimens: (a) SS-SiC, (b) HP-SiC, (c) F-Si_3N_4, (d) M-Si_3N_4, (e) C-Si_3N_4

generate dynamic deformation of specimens. Specimens were clamped with auxiliary fixture to avoid shattering of specimen during impact loading. Explosive indentation was performed using the electric bridge wire (EBW) detonator. The small explosive detonator encased in a stainless steel (150 μm thickness) cylinder and 5 mm diameter bottom plate. Ignition of EBW detonator propels the bottom stainless steel plate toward specimen having 1 km/s velocity to simulate dynamic indentation. After testing, the specimens were separated, cleaned, and the side-view of subsurface was observed with the optical microscopy with interference contrast and scanning electron microscopy with backscattered electron mode.

Observation of subsurface damages

Optical and scanning electron microscopy was used to observe subsurface damage of indented specimens. Indented specimens were gold coated for viewing in Normarski interference optical microscope which emphasizes the feature of micro-cracking. Observation of the subsurface damage zone in higher magnification was conducted by SEM. The subsurface damage of specimens was clearly shown in back-scattered image mode and inter or trans-granular fracture mode was observed in each specimen.

Dual focused ion beam milling (FEI, Netherlands) was adopted to prepare the specimens for TEM observation. TEM sample preparation with focused ion beam inhibits the unnecessary damage evolution during the sample grinding and thinning. This technique so enables observation of intrinsic fracture mode of dynamically indented specimens. The size of prepared TEM sample was 10 μm x 5 μm in rectangular shape with ~100 nm thickness. Transmission electron microscopy was performed on FIB-assisted specimens.

RESULTS AND DISCUSSION

Static and dynamic indentation

In Figure 2, subsurface damage of statically indented specimen was shown. Observed damage patterns are good agreements with general trends that the transition from fracture-dominated mode to deformation-dominated mode occurs with increasing heterogeneity of grain size and shape, ratio of abnormal and normal grains. As seen in Fig. 2, cone cracking is dominant

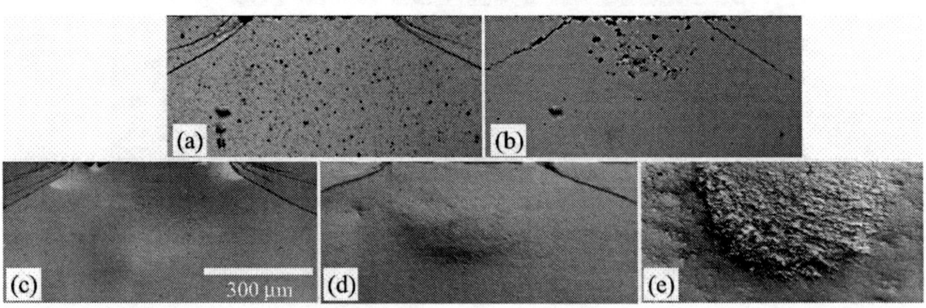

Figure 2 Static Hertzian indentation damage of Si_3N_4: (a) SS-SiC, (b) HP-SiC, (c) F-Si_3N_4, (d) M-Si_3N_4, (e) C-Si_3N_4

in SS-SiC and F-Si_3N_4, quasi-plasticity in C-Si_3N_4, and combination of cone cracking and quasi-plasticity in HP-SiC and M-Si_3N_4 specimens. These results mean that brittle fracture prevails in isotropic microstructure and quasi-plastic fracture in hetero-microstructure with liquid grain boundaries.

Figure 3 shows side-views of explosively indented specimens. The radial and lateral cracks were distributed in entire region of specimen, accompanying with the fragmentation and spallation. The compressive shock wave was propagated through the specimens and reflected at the edge of sample or interface between specimen and confinement metal blocks. These macroscopic cracking and spallation events seem to be generated after shock damage because the interference of propagated and reflected waves breaks the specimen into regular shaped fragmentation. It is still true that macroscopic cracks exist in all specimens and similar trends of crack pattern were seen independently on materials and microstructure. These fracture behavior is considered as a structural effect related to specimen dimension and test configuration so that this damage mode can be avoided with proper impedance matching design of confinement.

A particular characteristic of dynamic indentation damages is quasi-plastic damage mode in the specimens: HP-SiC, M-Si_3N_4, and C-Si_3N_4. The distinctive quasi-plastic damage pattern is pronounced, similar to the result of static indentation which is quasi-plastic damage prevails in HP, M, and C specimens. Shear faults or micro-fractures play an important role in quasi-plastic deformation in static indentation and these criteria seems to coincide with the damage behavior in dynamic indentation. The quasi-plastic damage zone of static indentation can be described as a cloud of many localized micro-cracks. In the HP-SiC specimen, the quasi-plastic damage zone is strongly distinguished from the region far from directly shock indented area, compared with the other materials. For silicon nitride specimens, quasi-plastic deformation zone area is increased

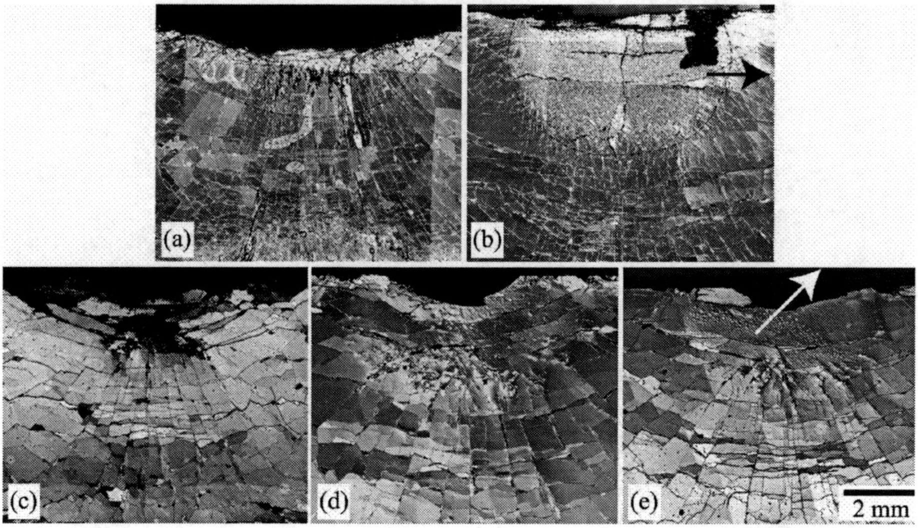

Figure 3 Explosive indentation damage of Si_3N_4: (a) SS-SiC, (b) HP-SiC, (c) F-Si_3N_4, (d) M-Si_3N_4, (e) C-Si_3N_4

with increasing amount of β elongated grain through the sequence F→M→C, however all specimens contains partially or fully developed quasi-plastic deformation zone. This dynamic fracture response of silicon nitride cannot be explained only with fracture mechanism of static indentation. The area of quasi-plastic deformed zone coincide with the static indentation results/ Fine grain silicon nitride shows no quasi-plastic deformation in static indentation but partially shows in dynamic indentation. It is unclear at this moment whether the other failure mechanism will be introduced especially in the fine grain ceramics with week grain boundary phase, such as F-silicon nitride.

Damage modes in static and dynamic indentation

Consider two silicon carbide with different microstructure: HP-SiC contains thin glassy phase in grain boundary and SS-SiC grain boundary with no glassy phase. These microstructural differences cause different fracture mode: quasi-plastic vs brittle damage modes. During dynamic compression, the weak glassy grain boundary of HP-SiC tends to fail easily, but grain boundary of SS-SiC is strong enough to endure the shock compression. Defeating weak grain boundary was also reported in Shih et al. and Lundberg et al.'s early studies.[10,11] In Fig. 4 (a) and (b), scanning electron micrographs with backscattered mode show the fracture mode of SS-SiC and HP-SiC. Intergranular and transgranular fracture modes are clearly shown in the observation of SS-SiC and HP-SiC specimen, respectively.

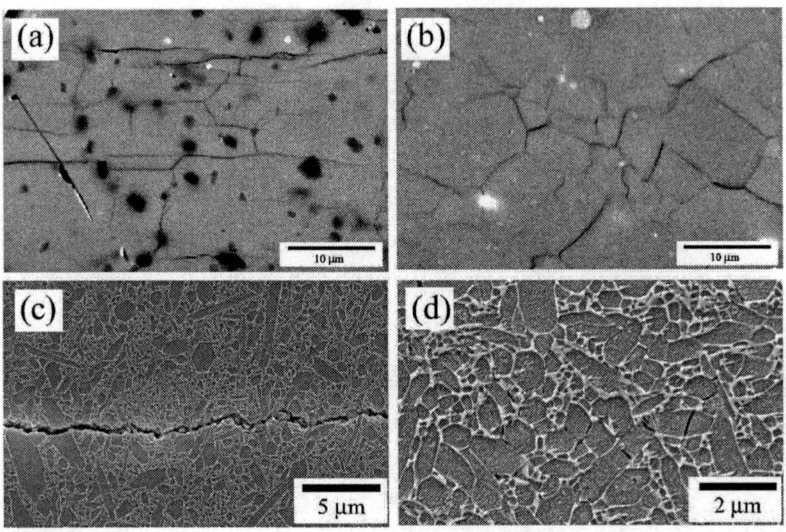

Figure 4 SEM micrographs of brittle and quasi-plastic deformation zone, BSE: (a) transgranular failure of SS-SiC, (b) intergranular failure of HP-SiC, (c) transgranular failure of F-Si_3N_4 (d) intergranular failure of C-Si_3N_4

Three grades of silicon nitride exhibit a transition from fracture-dominated to quasi-plasticity-dominated damage with variation of microstructure in static indentation experiment. In static indentation of Lee et al.'s work, this trend was explained with microstructural heterogeneity associated with the amount of elongated β grain.[12] This explanation can not be applied to the results of our explosive indentation experiments. High magnification observation of damaged zone in silicon nitride shows quasi-plastic failure mode in all three specimens. Even though three grades of silicon nitrides shows different strength, hardness, and fracture toughness, no significant change was found in dynamic indentation. The disagreement between static and dynamic indentation results can be explained with structure of grain boundary. In case of silicon nitride, all specimens contain almost same amount of weak liquid phase in grain boundary. It is considered that liquid grain boundary phase is primary factor for determining fracture mode during dynamic indentation.

In Fig. 5, TEM micrographs of dynamic indented HP-SiC and M-Si_3N_4 were shown. The TEM bright field images clearly indicate the intergranular fracture of HP-SiC and M-Si_3N_4. The shear fault and micro-cracking along the grain boundary phase is clearly identified. Grain boundary debonding seems to be a primary source of damage evolution of dynamically indented silicon carbide and silicon nitride.

CONCLUSIONS

Two grades of SiC and three grade of Si_3N_4 were indented in static and dynamic by sphere indenter and explosive detonator, respectively, to study the damage evolution. Hot pressed SiC and coarse grained Si_3N_4 show quasi-plastic deformation under static sphere indentation, whereas solids-state sintered and fine-grained Si_3N_4 show brittle fracture behavior. In case of dynamic indentation, SiC and Si_3N_4 containing liquid grain boundary phase show intergranular fracture mode which is similar to quasi-plasticity in static indentation. Failure

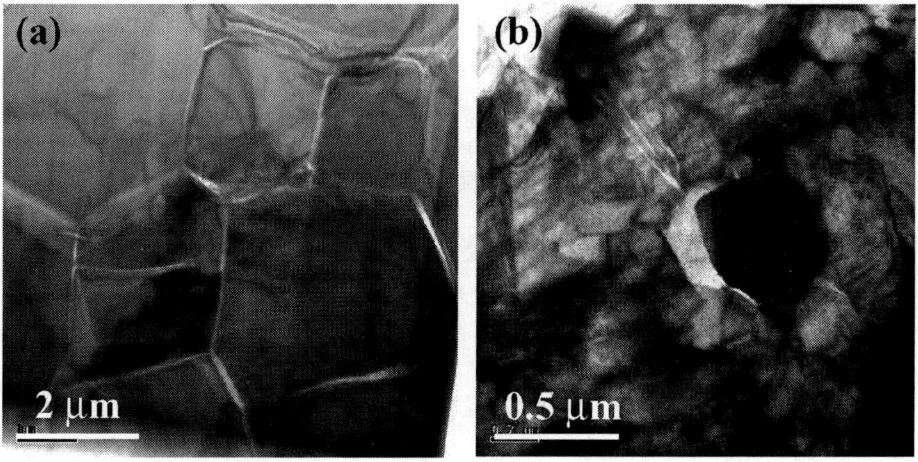

Figure 5 TEM micrographs of quasi-plastic deformation zone, bright field image: (a) intergranular failure of HP-SiC, (b) intergranular failure of C-Si_3N_4.

initiation mechanism of dynamic indentation was mainly from the defeating glass grain boundary which results quasi-plastic deformation. Intergranular fracture mode was observed in SEM and TEM observation at high magnification. It is considered that the controlling the weakness of grain boundary of ceramics is key factor to determine resistance to dynamic fracture.

REFERENCES

[1]David W. Richerson, Douglas W. Freitag, Technical Report DOE/ORO 2076 1-25 (1998)

[2]P. F. Becher, S. L. Hwang, H. T. Lin, T. N. Tiegs, "Tailoring of Mechanical Properties of Si3N4 Ceramics", *Kluwer Academic Publisher*, (1994)

[3]B. R. Lawn, "Indentation of Ceramics with Spheres: A Century After Hertz", *J. Am. Ceram. Soc.*, 81 1977 (1998)

[4]S.K. Lee, K. S. Lee, B. R. Lawn and D. K. Kim, "Effect of Starting Powder on Damage Resistance of Silicon Nitrides", *J. Am. Ceram. Soc.*, 81 2106 (1998)

[5]M. A. Meyers, Dynamic Behavior of Materials, *A Wiley Interscience Publication*, 1994.

[6]D. K. Kim, C-S Lee and Y-G. Kim, "Dynamic Indentation Damage of Ceramics", *Proceedings of the Symposium on Ceramic Armor Materials by Design.*, PACRIM IV, pp. 429-440 (2001)

[7]J. H. Kim, Y-G. Kim, and D. K. Kim, C-S. Lee, D-T. Chung, C. W. Kim, J. H. Choi, and S-N Chang, "Controlled Explosive Indentation on Ceramics", *Proceedings of the Ceramic and Ceramic Armor Systems Symposium*, the 105th the American Ceramic Society Meeting, pp. 93-104 (2003)

[8]J. H. Kim, Y-G. Kim, D. K. Kim, K. S. Lee, and S-N Chang, "Static and Dynamic Indentation Damage in SiC and Si_3N_4", *Key Eng. Mater.*, 287 410-415 (2005)

[9]F. Guiberteau, N. P. Padture, B. R. Lawn, "Effect of Grain Size on Hertzian Contact Damage in Al2O3", *J. Am. Ceram. Soc.*, 77 1825 (1994)

[10]C. J. Shih, M. A. Meyers, V. F. Nesterenko and S. J. Chen, "Damage Evolution in Dynamic Deformation of Silicon Carbide", *Acta. Mater.*, 48 2399 (2000)

[11]P. Lundberg, R. Renstrom, B. Lundberg, "Impact of Metallic Projectiles on Ceramic Targets: Transition Between Interface Defeat and Penetration", *Int J. Impact Engng.*, 24 259 (1988)

[12]D. K. Kim and C-S Lee, "Indentation Damage Behavior of Armor Ceramics", *Proceedings of the Symposium on Ceramic Armor Materials by Design*, PACRIM IV, pp. 429-440 (2001)

COMPRESSIVE FAILURE THRESHOLD OF BRITTLE MATERIALS

E.B. Zaretsky, V.E. Paris,
Department of Mechanical Engineering, Ben-Gurion University of the Negev, P.O.B. 653, Beer-Sheva 84105, Israel

G.I. Kanel, Institute for High Energy Densities, Russian Academy of Sciences, Izhorskaya 13/19, Moscow, 125412, Russia

A. Rajendran, U.S. Army Research Laboratory, ARO, RTP, NC 27709-2211, USA

ABSTRACT

Planar impact experiments widely used for determining dynamic strength of hard materials do not provide a possibility to obtain information about the mode, ductile or brittle, of their inelastic response or about thresholds conditions of their compressive failure. Using the controlled sample pre-stressing in the experiments (E. Zaretsky and G. Kanel, *APL*, 2002, 81, 119) allows one only to conclude whether the response of the studied material at the stress level of elastic limit is brittle or ductile leaving the problem of the material compressive failure threshold still unsolved.

The experimental/numerical technique for determination of compression failure threshold of brittle solids is presently suggested. The technique is based on measuring the velocity of the sample/window interface while the sample front surface is struck by curved copper impactor. The divergent stress flow produced by such loading results in the incipient compressive failure of the sample and, respectively, in the appearance of the failure signatures at the measured velocity profile. Reproducing these signatures with aid of AUTODYN 2D commercial computer code including the constitutive and failure description of the studied material yields the values of the principal stress tensor components corresponding to the brittle failure initiation. The locus of these values in the principal stress space is the material compressive failure surface.

INTRODUCTION

Inelastic deformation, fracture, and fragmentation caused by impact or explosion occur under three-dimensional loading conditions, where at least one stress component is compressive. Description of inelastic deformation, cracking, and fragmentation of targets made of brittle material under such loading requires the knowledge of the material dynamic damage threshold as a function of the stress state. Since the mechanisms of brittle or ductile deformation are different the criteria of the onset of the inelastic deformation are also different for ductile and brittle materials. In the case of ductile materials the von Mises yield criterion is commonly accepted and gives satisfactory results. In the case of brittle materials the situation is not so unambiguous. The use of von Mises criterion does not fits the strength data obtained after indentation hardness measurements with the strength data based on Hugoniot Elastic Limit σ_{HEL} values obtained after planar impact experiments. The Griffith's criterion of compressive fracture fits the strength data better [1, 2] but the strength calculated with aid of this criterion from experimental values of σ_{HEL} seems underestimated [3]. Experimental determination of the failure threshold of hard brittle ceramics [4 - 7] meets multiple difficulties. It has been shown [5] that a transition from the brittle behavior at ambient pressure to the ductile one at the presence of high enough lateral compressive stress occurs in many ceramics and rocks. Since the mechanisms of inelastic deformation are different for these two, ductile and brittle, modes of material behavior the conditions of the brittle-ductile transition and the criteria governing the material response before and after the transition are crucial for predicting the compressive response of ceramics.

Planar impact experiments are widely used for obtaining quantitative information about dynamic inelastic response of materials. The 1-D strain conditions of such experiment provide accurately defined material flow in the studied sample facilitating the interpretation of the experimental results. In the real impact, however, the ratio σ_2/σ_1 between principal transversal σ_2 and longitudinal σ_1 stress may vary within wide limits whereas under planar impact loading the ratio stays constant: below the HEL it is equal to $\sigma_2/\sigma_1 = \nu/(1-\nu)$ where ν is the Poisson's ratio. In addition, the upper limit of the material elastic response, σ_{HEL}, detected in planar impact experiment cannot be attributed unambiguously to the material transition from the elastic behavior to either ductile or brittle inelastic one.

Recently suggested experimental technique [3, 6] allows one to reveal the mode of inelastic (ductile or brittle) material behavior under shock wave loading. The technique uses different sensitivity of the onset of ductile yielding or brittle failure to the small variation of transversal stress. The planar impact experiments were performed with the disk-like alumina [6] and boron carbide [3] samples pre-stressed (radial compressive stress was of about 0.3 GPa) by a shrink-fit steel ring, so that the sum radial compression at HEL was varied in experiments with alumina from 1.6 GPa (free samples) up to 1.9 GPa (pre-stressed samples). For the radial pressure above 1.6 GPa the inelastic response of the alumina samples was found ductile while, as shown by Heard and Cline [5], for the radial compression below 1.25 GPa the behavior of alumina is still brittle. Thus, the brittle-ductile transition in alumina should be expected somewhere between these two values of the radial compression. In similar experiments [3] with boron carbide no hint on such transition was detected under radial compression of about 4 GPa; up to this value the inelastic response of boron carbide stays apparently brittle. It is plausible to assume that such material, whose response under planar impact loading was proved to be brittle, will show the brittle behavior under arbitrary impact loading. In contrast to the boron carbide the information available after planar impact with alumina is insufficient for predicting the alumina response under arbitrary impact loading: we need to know the conditions of the brittle-ductile transition in order to describe the material behavior. It would be, therefore, desirable to expand the capabilities of shock-wave experiments in order to obtain measurements over a wider range of stress states characterized by different ratios of longitudinal to transversal stress.

For obtaining such information the use of impact experiments with divergent material flow seems reasonable: the longitudinal-to-transversal stress ratio may be varied in such experiments within wide limits. During the divergent compression, the stressed state of tested material may correspond to either compressive failure threshold or ductile yielding.

The main objective of this work was to extend the shock-wave testing capabilities towards the divergent impact loading of studied materials, most of all the brittle ones, in order to vary the ratio between longitudinal and transversal stress and to obtain experimental information about conditions of brittle-ductile transitions. The divergent loading was realized using impactors with spherical impact surface of relatively large (200 mm to 600 mm) radius of curvature. It will be shown that such impactor generates in a high-impedance target plate a divergent material flow whose symmetry in the vicinity of the impact axis is close to the spherical one. The stress states corresponding to the material failure and post-failure behavior were determined by means of

VISAR[8] measurements of generated waveforms followed by the AUTODYN™-2D[9] simulations of the experiments.

DIVERGENT FLOW IN CLOSE-TO-PLANAR IMPACT

In the planar impact experiment (1-D strain loading) the ratio of the longitudinal to radial stress stays constant below HEL:

$$\sigma_2/\sigma_1 = \nu/(1-\nu). \tag{1}$$

The measured value of the Poisson ratio of the alumina studied in the present work is ν= 0.23. As was shown[6], the response of alumina to the planar impact loading above HEL is plastic yielding. Assuming for the alumina elastic-perfect plastic behavior with strength Y_{duct} = 4.3 GPa[4] we can define on the principal stress plane (σ_1,σ_2) the region of interest (Fig. 1a). It is the triangle *ABC* bounded by the σ_1 axis (σ_2 = 0) and two straight lines: the bottom one, $\sigma_1 = \sigma_2 (1 - \nu)/\nu$, and the upper one, $\sigma_1 = Y_{duct} + \sigma_2$, related to the von Mises yield condition. At the point of the intersection of the lines (HEL, point C) the longitudinal and the radial stress are equal to σ_1 = 6.10 GPa and σ_2 = 1.82 GPa, respectively.

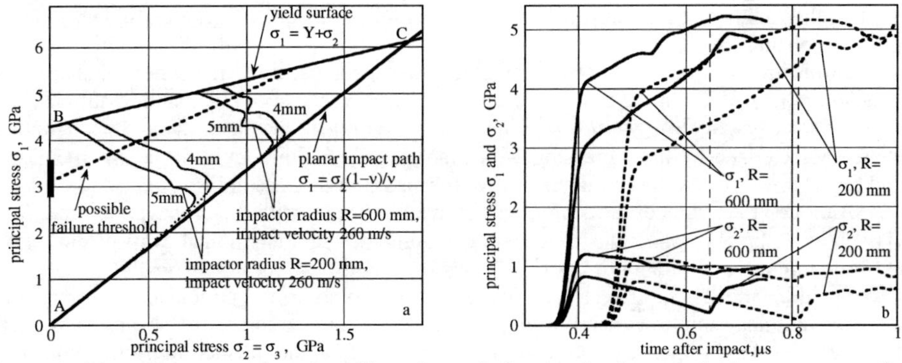

Figure 1. Planar impact path and yield surface of alumina in the principal stress space. Solid rectangle corresponds to the literature data on the alumina strength measured by rod impact technique[4, 7] (a). Stress histories $\sigma_1(t)$ and $\sigma_2(t)$ on the sample axis corresponding to the material trajectories shown in Fig 1a. Solid and dashed curves show the stress profiles obtained at 4 and 5 mm from the impacted sample surface, respectively. Dashed straight lines show the instant of the shear wave arrival causing the fast increase of the radial stress (b).

According to Heard and Cline data[5] alumina stays brittle for radial stress σ_2 < 1.25 GPa. That means that for the radial stress less than 1.25 GPa the states corresponding to the threshold of the brittle alumina failure in compression should be located on the principal stress plane below the von Mises line $\sigma_1 = Y_{duct} + \sigma_2$. Such hypothetical threshold is shown in Fig. 1a by dashed line. As apparent from Fig. 1a that the material states at the threshold cannot be achieved under planar impact loading.

Analyzing particle trajectories of the spherically symmetric flow generated in an elastic body with spherical cavity Grady [10] show that the particle trajectories in the principal stress space depart from the planar impact path $\sigma_1 = \sigma_2\ (1 - \nu)/\nu$ toward smaller σ_2 values. Although it is evident that generating the spherical wave is capable of providing the required stress states, the experimental implementation of such loading geometry is hardly possible. Loading configuration close to the planar impact but using impactors with spherical front surface is capable of providing the stress states close to those provided by perfect spherical loading. Such possibility was verified in preliminary AUTODYN-based numerical simulations of close-to-planar impact experiments. In the simulations special AUTODYN tools capable of matching the curved boundaries with the square mesh of the problem were used. The mesh size of the preliminary simulations was 33 micron. The copper and alumina properties available from the AUTODYN material library were employed in these simulations, and the shock parameters (pressure, stress, particle velocity) were recorded at the points located on the sample axis at 1, 2, 3, 4, and 5 mm from the impact surface. Corresponding material trajectories $\sigma_1(\sigma_2)$ obtained in the AUTODYN runs for the sample points located on the impactor-sample at the distance 4 and 5 mm from the impacted sample surface are shown in Fig. 1a. Figure 1b shows the stress histories $\sigma_1(t)$ and $\sigma_2(t)$ at the same (4 and 5 mm) points of the alumina target. The simulations show that convex copper impactor of radius R_i having velocity V_i generates in the alumina sample a spherical longitudinal wave having initial radius of curvature close to $R_{sl0} = R_iV_i/C_l$, where C_l is the longitudinal speed of sound in alumina. Note, that the radius of curvature of the wave front, R_{sl0}, is much smaller than radius R_i of the impactor curvature.

Surplus result of such impact is the simultaneous generation of the spherical shear wave of smaller, $R_{ss0} = R_iV_i/C_s$, initial radius. Here C_s is the transversal speed of sound in alumina. The shear wave changes the flow geometry: as apparent from Fig. 1b, the flow behind the longitudinal wave is characterized by the increasing with time stress difference σ_1 - σ_2 but this increase is halted by the arrival of the spherical shear wave (dashed straight lines in Fig. 1b).

Although the radius of the longitudinal wave is growing with the propagation time, $R_{sl0} = R_{sl0}[1+(\ C_l^2 - V_i^2)t/\ R_iV_i]$, and the center of curvature of the longitudinal wave front moves backward with velocity of about $(V_i^2/\ C_l^2)C_l$ this departure from ideal spherical symmetry does not change dramatically the trajectories of the material points in the principal stress space: the flow in the alumina samples is nearly spherical between the longitudinal and shear waves within ±30 degrees from the impact axis. It is apparent from Fig. 1a, that in such impact geometry the brittle failure characterized by proper combination of the principal stress may be initiated in the different cross-sections of the sample.

AUTODYN simulations provide possibility to estimate the influence of the departure of the impact geometry from the designed one. This is extremely important when the impactor-sample contact point and the VISAR observation point do not coincide with the centers of the sample front and rear surfaces. Preliminary simulations show that up to some 2-mm non-concentricity its influence on the VISAR records may be ignored. The presence of failure in the sample makes the concentricity requirement more rigorous, so special attention was paid to this issue during implementation of the experimental part of the work. As well, the influence of possible inaccuracies of turnery (during preparation of spherical impactors) on the experimental records (deviation from the spherical shape, the uncertainties of the impactor radius) was studied with aid of multiple AUTODYN runs.

Use of curved impactors gives rise to several technical problems. One of them leads to impossibility of collecting the VISAR signal from the free surface of the studied sample. The longitudinal stress at the free sample surface is always zero while the radial stress becomes tensile just immediately after arrival of the spherical shock at the surface. This may cause the tensile fracture of the surface sample layer prior to its failure in compression. The use of sapphire window and, respectively, the VISAR record of the velocity of the alumina-sapphire interface solve the problem.

Another problem, caused by use of the curved flyer plates is related to the simultaneous generation of both longitudinal and shear waves at the impacted region. The wishful divergent flow exists behind the longitudinal wave front. The shear wave partially restores the parallelism of the particle velocity vectors dictated by the conservation of momentum but cancels the flow divergence. In the case of 5-mm alumina samples the time interval during which the information related to the divergent flow arrives at the alumina-sapphire interface does not exceeds 330 ns.

It should be noted that the initiation of the failure is separated both in time and space from the impactor tip touching the sample surface: at the very instant of the touch the flow at the contact point is not divergent (two waves, the longitudinal one and the shear one, are not separated). The flow departure from parallelism (and, as well, the stress difference $\sigma_1 - \sigma_2$) is growing with time and with the propagation distance and eliminated just behind the shear wave front. Thus, the failure initiation is expected in the immediate region ahead of the shear wave while the start of the failure should be accompanied by the unloading signal propagating towards the VISAR observation point with the longitudinal speed of sound. The appearance of the unloading signal on the recorded velocity profile may be used for estimation of the location of the failure initiation point. The stress difference required for this initiation may be achieved after some delay time (with respect to the impact instance) increasing with the increase of the impactor radius. This yields a limitation on the increase of the impactor radius: the failure initiation should occur inside the sample.

The short observation time terminated by the arrival of the shear wave may be even shorter because of the possibility of the development of the hydrodynamic instability at the impactor-sample contact surface. The velocity $U=V_i(R_i-V_it)/(2R_iV_it-V_i^2t)^{1/2}$ of the propagation of the border of the impactor-sample contact surface decreases continuously with time t since the contact between the tip of the impactor and the surface of the sample took place. At the time instant $t = RV/2C_l^2$ after the impact the velocity U should become smaller than speed of sound C_l in the sample. Such situation, $U < C_l$, may result in the development of the hydrodynamic instability at the contact border accompanied by the wave formation (corrugation of the contact interface) similar to that taking place during explosive welding [11]. When $RV < 2 \cdot 330\ \text{ns} \cdot C_l^2 \approx 74$ the disturbance caused by the development of the instability will arrive at the alumina-sapphire interface prior to the arrival of the shear wave disturbance. Fortunately, the condition $U < C_l$ is necessary but not sufficient for the wave formation [11]: the wave formation requires some unloading signal for its triggering. Using copper impactors with thickness larger than 2 mm prevents the wave formation during some about 1 μs after the impact.

MATERIALS AND EXPERIMENTAL TECHNIQUE

The studied samples were 25-mm diameter and 5-mm thickness alumina disks. The disks were precisely (2-μ tolerance, 0.1 mrad parallelism) cut from the rod of 99-% purity hot-pressed alumina (Microceramica Ltd., Karmiel, Israel). The density ρ_0 of the samples was determined by the liquid displacement technique using distilled water. The longitudinal C_l and the shear C_s speeds of sound were measured by ultrasonic pulse-echo technique using 5-MHz probes. The measured properties of the alumina samples including Poisson's ratio $\nu = [(C_l/C_s)^2-2]/[2(C_l/C_s)^2-2]$ are listed in the Tab. 1.

The geometry of impact experiment with spherical impactors is shown schematically in Fig. 2.

Table 1. Physical properties of the studied alumina samples.

sample	ρ_0, g/cm^3	thickn., mm	C_l, m/sec	C_s, m/sec	Poisson's ratio, ν	G, Gpa
1*	3.896	4.987	10566	6200	0.237	150
2	3.907	5.029	10550	6240	0.232	151
3*	3.906	5.047	10550	6230	0.233	151
4	3.900	5.010	10620	6240	0.237	152
5	3.912	5.010	10560	6240	0.231	152
6	3.901	5.063	10550	6200	0.235	150
7	3.906	5.048	10650	6260	0.236	153
8	3.899	5.061	10630	6210	0.241	150
9	3.896	5.058	10630	6220	0.240	151
average	3.902	5.034	10580	6230	0.235	151
uncert.**	0.006	0.003***	50	19	0.004	2

*Samples used in planar shots. **Uncertainty of a property includes both the inaccuracy of the property measurements and the standard deviation from the property mean value. ***The inaccuracy of the thickness measurement only.

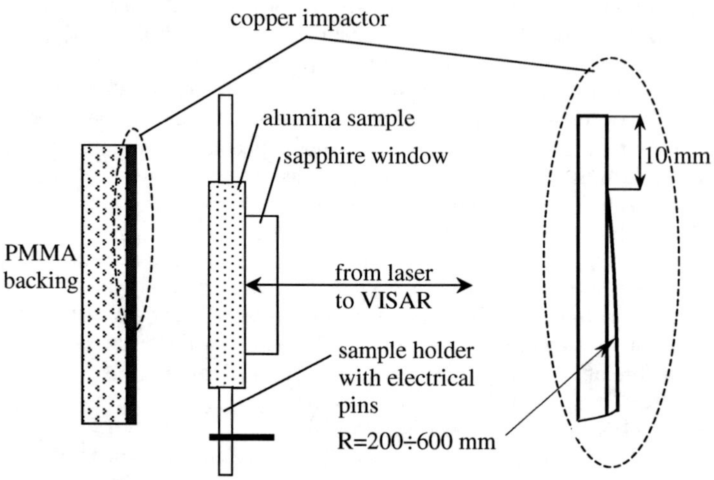

Figure 2. Close-to-planar loading of the ceramic sample by curved copper impactor. In the experiments with plane impactors the VISAR signal was collected directly from the free sample surface.

The shots were performed with 58-mm, 4.2-m length gas gun of the Laboratory of Dynamic Behavior of Materials at Ben Gurion University [6]. Convex impactors were prepared by the following procedure: The discs of 56-mm diameter cut of the 5-mm OFNC copper sheet were glued to the front surface of 56-mm diameter, 15-mm thickness PMMA backing disks which, in turn, were glued to the front edge of the hollow aluminum projectile. Further machining of the copper head of the assembly by precise CNC (the shape shown with magnification in Fig. 2) provides the projectile-impactor concentricity. Uncertainty of the impactor surface curvature did not exceed +30 mm for impactor radius R = 200 mm and +90 mm for impactor radius R = 600 mm for the impactor part close to the projectile axis. Shortening of the velocity and misalignment pins was produced by 10-mm impactor shoulders, Fig. 2. In all shots the shoulder-sample misalignment did not exceed 2 mrad.

One surface of the sapphire windows of 6-mm thickness and 20-mm diameter, Fig.2, was gold sputtered in order to provide reflectivity of the sample window interface sufficient for the VISAR operations. The thickness of the gold layer was about 0.4 μm. The sputtered window surface was glued to the rear surface of the alumina samples. In the experiments with curved impactors the velocity of the alumina-sapphire interface was monitored by VISAR having interferometer constant 216 m/sec per fringe. Prior to each shot special efforts were made in order to provide collinearity of the projectile and the sample centers and of the laser spot at the sample-window interface so, that the non-concentricity did not exceed 0.5 mm.

Planar impact experiments with the same type allumina samples were performed using projectile made of 2-mm copper disc glued to 15-mm PMMA backing. In these shots the velocity of the free surface of alumina samples were recorded by VISAR with the same interferometer constant. The impactor-sample misalignment in these shots did not exceed 0.5 mrad.

RESULTS OF THE EXPERIMENTS

VISAR records of two alumina samples tested in planar impact experiments are shown in Fig. 3a. The sample free surface velocity at the alumina Hugoniot Elastic Limit is equal to u_{fs} = 283±3 m/s. This yields for the alumina HEL $\sigma_{HEL} = 0.5\rho_0 C_L\, u_{fsHEL}$ = 5.84±0.09 GPa and for the alumina strength $Y = (1-2\nu)/(1-\nu)$ = 4.1±0.2 GPa. Although the measured values differ slightly from those used for design of the experiments with the curved impactors, σ_{HEL} = 6.10 GPa and Y = 4.3 GPa, there was no need in the correction of the designed impact geometry.

Typical alumina-sapphire interface velocity profiles for different impact velocities and 600-mm radius of the impactor curvature are shown in Fig. 3b. The fast initial ramp of the interface velocity corresponding to the material motion, in the principal stress space, along the line $\sigma_1 = \sigma_2 (1 - \nu)/\nu$ (see. Fig. 1a) is followed by the slow velocity increase corresponding to the divergent flow accompanied by the increase of the stress difference $\sigma_1 - \sigma_2$. The distinct failure signatures (shown by arrows) are apparent on both profiles. At the absence of the failure the interface velocity should continue to grow up to the instant of the arrival of the shear wave (dashed line *I* corresponds to the arrival of the shear wave propagating through undamaged alumina) at the interface. Using the derivative of sapphire refractive index on compression, k_0=0.7864 [12], the VISAR-measured interface velocities were corrected for time interval between the arrival of the shock signal at the sample-sapphire interface and the arrival of signal at the sapphire free surface (dashed line *II*). The difference between literature data on sapphire refractive index n_0 = 1.771763 [12] (532 nm) and that for actual VISAR light (λ = 514) nm was neglected as well as the influence of the wave curvature on the index correction. The latter is justified by negligibly small difference $\mu_0 - k_0 = n_0 - 1 - k_0$ to which the curvature correction is proportional [13].

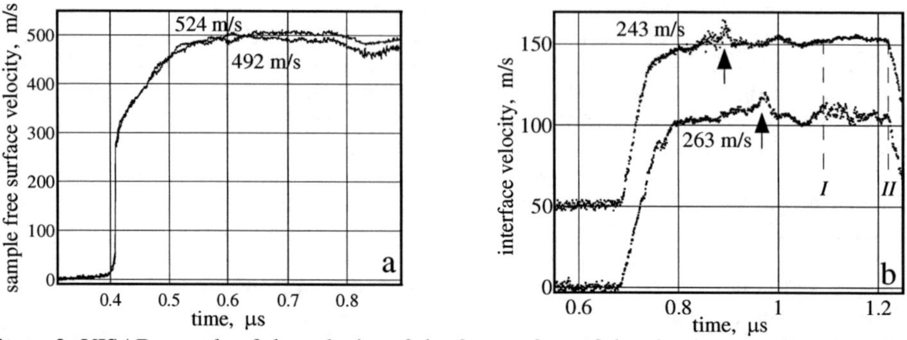

Figure 3. VISAR records of the velocity of the free surface of the alumina samples after planar impact loading - (a). Typical VISAR records of the velocity of alumina-sapphire interface obtained in the impact loading of the plane-parallel alumina samples by curved copper impactors. The arrows correspond to the arrival of the failure initiation signal at the interface. Dashed lines correspond to the arrival of the shear wave at the interface (*I*), and to the arrival of the longitudinal wave at the free sapphire surface (*II*) - (b).

Another type of the velocity profile having distinct but not a single failure signature (marked with arrows) is shown in Fig. 4a.

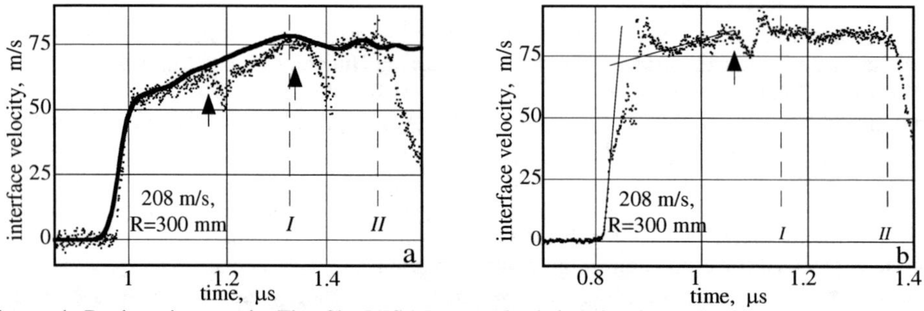

Figure 4. Designations as in Fig. 3b. VISAR record of the alumina-sapphire interface velocity with two failure signatures (arrows). Solid line corresponds to the computer simulation of the impact. Right arrow corresponds to the failure event taking place at some 2.2 mm from the sample axis (a). Departure of the recorded VISAR profile from its expected shape (straight lines) caused by the impactor surface defect (b).

The interface velocity profile obtained after AUTODYN simulation of this impact experiment (without failure) is shown in Fig. 4a by solid line in order to emphasize the instant of the arrival of the shear wave at the interface (dashed line *I*) halting the stress difference (and the interface velocity) growth. The first (left) velocity downfall corresponds to the rapid fracture (rapid departure from the computed fracture-less curve) initiated on the alumina sample axis at the distance of about 0.8 mm from the sample-sapphire interface (of about 4.2 mm from the impact surface). The second velocity downfall seems corresponds to the failure initiation aside the sample axis; at the axis the stress difference is just cancelled by the arrival of the shear wave. Assuming that the spherical surface with 4.2 mm radius is the locus of possible failure initiations (the same level of the stress difference is achieved) the failure corresponding to the second velocity downfall should be initiated at the distance of about 2.2 mm from the sample axis.

Non-ideality of the impactor spherical surface also leads to the appearance of additional failure-like signatures on the velocity profile as shown in Fig. 4b. During time interval of about 100 ns the shape of the initial part of the velocity profile differs from the expected (shown by thin solid lines). Accounting in that the perturbation of the profile starts some 20 ns after arrival of the shock at the sample-window interface it is plausible to assume that the disturbance is caused by small lug of the impactor surface close to the impactor tip (the lug located directly at the tip will disturb the velocity profile prior to the start of the velocity ramp). AUTODYN simulations show that the triangle lug with the 30-micron base and the same height (always possible under turning) is capable of producing such profile perturbation. Simulations show that the fracture starts at the close vicinity of the point of the lug-sample collision but stopped after flattening the lug.

NUMERICAL SIMULATIONS. ALUMINA FAILURE THRESHOLD

As mentioned previously, both the design of the experiments and the numerical simulations of the results of the actual experiments with divergent shock compression were performed with 2-D axisymmetrical Lagrangian solver of commercial AUTODYN 2-D code [9]. The mesh size in all simulations with failure was 100 μm for all parts of the system. As mentioned above, at the design stage the standard AUTODYN library was used for description of the local material properties; elastic-perfect plastic response was accepted for all three materials: copper, alumina, and sapphire. The library properties of alumina were modified according to the measured average values of the studied samples. In particular, the alumina yield strength was chosen equal to Y_0 = 4.3 GPa. In the case of numerical simulation of actual experiments some minor deviations of alumina Y_0 from the chosen value were permitted. The alumina ductile response was described by von Mises yield criterion

$$(\sigma_1 - \sigma_2)^2 \le Y_0^2 \text{ or } \sigma_{1y} = Y_0 + \sigma_2, \tag{2}$$

while a simple description of the alumina brittle compressive failure threshold, suggested by Ashby and Sammis [14] and having similar to Eq. (2) linear form

$$\sigma_{1\text{br}} = \sigma_{0\text{br}} + f \cdot \sigma_2 \tag{3}$$

was added. Here and in the following the stress is positive in compression. The Eq. (3) corresponds to the possible compressive failure threshold of Fig.1a. Crossing or touching the failure threshold line by the material trajectory in the principal stress state triggers the material compressive failure accompanied by the increase of the material damage parameter $D(\varepsilon_{\text{ie}})$ varied between zero (no failure) and one (complete failure) and related to the inelastic (produced by the equivalent stress excess with respect to the failure threshold) strain ε_{ie} via linear equation

$$\frac{dD}{d\varepsilon_{ie}} = k_D. \tag{4}$$

The coefficient k_D characterizes the damage kinetics. The description of the post-failure behavior of alumina was based on the Mohr-Coulomb criterion, attributing the material ability to withstand the shearing because of the friction between the particles of the comminuted material. The comminuted material trajectory in the principal stress space is also described by the linear equation similar to (2) and (3):

$$\sigma_{1\text{fr}} = \sigma_{0\text{fr}} + c \cdot \sigma_2. \tag{5}$$

The departure of the constant c from 1 corresponds to the pressure-dependent friction stress $\tau_{\text{fr}} = (\sigma_1 - \sigma_2)/2$. When $c = 1$ the friction is independent on pressure and equal to $\tau_{\text{fr}} = \sigma_{0\text{fr}}/2$. When the failure threshold has been achieved and the failure has been initiated the local failure threshold σ_D departs from its initial value described by Eq. (3) and moves towards the trajectory of the comminuted material, Eq. (5):

$$\sigma_D = D\sigma_{1\text{br}} + (1 - D)\sigma_{1\text{fr}}. \tag{6}$$

If the local equivalent stress exceeds the local failure threshold it relaxes towards the threshold value σ_D during the relaxation time τ_r [15]. The value of the relaxation time of τ_r = 10 ns corresponding to the traveling of acoustic signal for a distance of one numerical cell, 100 μ, was chosen. The described constitutive behavior of alumina, Eqs. (3 - 6), was added to the corresponding AUTODYN problem in a form of FORTRAN subroutine.

Accounting in that even in the simplest formulation the description of the alumina compressive failure requires the knowledge of five fitting parameters, $\sigma_{0\text{br}}$, f, $\sigma_{0\text{fr}}$, c, and k_D, the search of the parameters was conducted in three stages. At the first stage no failure subroutine was employed in the AUTODYN numerical simulations of the experiments, exactly, as it was done for design of the experiments. Such simulation yields sample-window interface velocity profiles for free-of-failure alumina samples (one of such velocity profiles is shown in Fig. 4a). As well, such simulation yields the time dependence of stress σ_1 and σ_2 for any sample point. Accounting in that the both the point x_f and the instant t_f of the failure initiation at the sample axis may be estimated from the time of arrival of the failure feature on the interface (left arrow in the Figs. 4a, b) the stress $\sigma_{1\text{br}}$ and $\sigma_{2\text{br}}$ at the point may be found from the simulated stress profiles $\sigma_1(t)$ and $\sigma_2(t)$ obtained at the point x_f. Such procedure repeated for several profiles yields the draft equation for compressive failure threshold $\sigma_{1\text{br}}(\sigma_{2\text{br}})$:

$$\sigma_1 = \sigma_{0\text{br}} + f \cdot \sigma_2, \tag{7}$$

where $\sigma_{0\text{br}}$ = 3.47 GPa and f = 1.68. Note, that the value of $\sigma_{0\text{br}}$ is in agreement with the alumina strength values found in the rod impact tests [4, 7]. These values of $\sigma_{0\text{br}}$ and f were used as the priming values at the second stage of the search of the alumina failure parameters. At this stage each experimental profile was simulated under following conditions:

(i) The brittle threshold parameters σ_{0br} and f for each shot should not differ strongly from their priming values.

(ii) The values of kinetic coefficient k_D should be similar for all shots.

(iii) The impact velocity in simulation should not differ from measured impact velocity for more than 2% (accuracy of the impact velocity measurements).

(iv) The impactor radius used in the simulation should not differ from the corresponding radius claimed by the mechanical treatment for more than the radius uncertainty. (see Section Materials and Experimental Technique of the paper).

(v) Since no primary information concerning the trajectory of the comminuted material, Eq. (5) was available the initial simulations of this stage were concentrated on the search of the σ_{0fr} value while the value of the slope c in Eq. (5) was taken equal to the slope f of the brittle threshold, Eq. (7). When the primary σ_{0fr} value for a given shot was found, the pair σ_{0fr}-c was optimized by trial and errors. After this procedure it was found that the values of c and f are really very close and the value 1.85±0.01 may be accepted for both slopes.

When the individual (for each shot) values of parameters were determined the third stage of the simulation procedure was performed. At this stage only minor, within 5-10%, departures of the values of the failure parameters from the average values obtained after the second stage were permitted. Table 2 shows the average values of the alumina failure parameters together with their values found for individual shots after several iterative simulations. Note, that the values of σ_{0br} of Tab. 2, 3.35 – 3.50 GPa, stay in satisfactory agreement with the alumina strength data, 2.8 – 3.6 GPa, obtained in rod impact experiments with alumina bars [4, 7].

Figure 5 allows comparing the fit of the experimental VISAR records by the curves obtained using the best-fit set of parameters, with that obtained using the set of average parameters (obviously, at the expense of the fitting quality) and with that obtained using the set of the average parameters but for VISAR observation point located 1 mm apart from the sample center. The latter curve justifies the measures performed for the sake of projectile-sample-laser spot collinearity. Figure 6 shows the influence of varying the failure parameters on the change of the failure feature with respect to that of the best-fit curve.

Table 2. Individual for each shot and average failure parameters.

SHOT*	imp. vel., m/s	Y_0, GPa	σ_{0br}, GPa	f	σ_{0fr}, GPa	c	τ_r, ns	k_D
300 (300)	210 (208)	4.30	3.35	1.86	1.30	1.86	10	300
380 (300)	232 (225)	4.30	3.50	1.86	1.30	1.86	10	200
660 (600)	263 (263)	4.38	3.50	1.86	1.30	1.86	10	270
600 (600)	250 (250)	4.30	3.35	1.86	1.30	1.86	10	200
680 (600)	250 (243)	4.30	3.40	1.86	1.30	1.86	10	200
380 (300)	270 (259)	4.00	3.53	1.86	1.30	1.86	10	300
average		4.26	3.44	1.86	1.30	1.86	10	245
std. dev.		0.13	0.08	-	-	-	-	50
280 (200)	244 (244)	4.40	3.85	1.86	1.70	1.86	10	250

*The impact experiments are marked by the impactor radius used in the simulations. The radius value given in the parentheses correspond to the impactor radii expected after turning.

It is apparent from Fig. 6 that employed failure description shows different sensitivity to the variation of different failure parameters: the effect of 10-% variations of σ_{0fr} is much less than that of 1.5-% variation of σ_{0br} (Fig 7a, b). As well, the model sensitivity with respect to the variation of the failure kinetics, k_D, Fig. 7c, is relatively low. Different sign of the variations of alumina yield strength, Y_0, results in the different effects (Fig. 7d): the increase of Y_0 has almost no effect on the failure behavior while the decrease of Y_0 (approaching the stress state at the failure site to the yield surface) leads to the deceleration of the failure or to its complete suppressing.

The difference between best-fit parameters and the average one (see Tab. 2) yield an estimation of the uncertainty of the determination of the parameters of the alumina compressive failure. Unfortunately it is impossible to separate contributions of the experimental and simulation inaccuracies from the influence of the natural scattering of alumina properties. Although the scattering of the measured densities and velocities of the alumina samples were found very small (0.1%, 0.4%, and 0.3% for density, longitudinal and shear speeds of sound, respectively), it is plausible to expect much higher scattering of the failure parameters caused by presence of the micro-cracks of different size at the failure initiation sites of different alumina samples. These expectation is confirmed by the dependence of the alumina strength on the sample size found in rod impact tests [4].

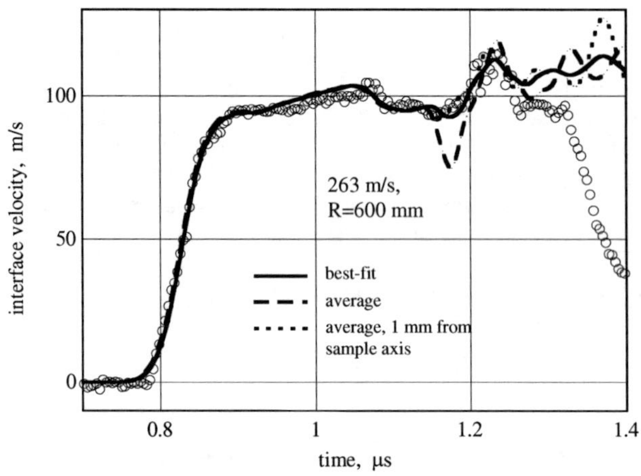

Figure 5. Experimental VISAR records (circles) of the alumina-sapphire interface velocity, and computer simulations of the impact with different values of failure parameters and different locations of the VISAR observation point.

The bottom row of Tab. 2 contains results of the simulation of the only shot performed with impactor having radius of curvature of about 200 mm. The failure parameters for this shot, especially σ_{0br} and σ_{0fr}, differ from both the average the values and from the values found for individual shots. This shot, with relatively small radius of impactor and moderate impact strength, was performed with the aim to cross the failure threshold surface as close as possible to the $\sigma_2 = 0$ axis (in principal stress space). This, however, results in the decrease of the product $RV \approx 49$

(see Section Divergent Flow in Close-to-Planar Impact of the paper) much below the critical value $RV = 74$. In such shot geometry the arrival of the hydrodynamic shear (corrugation) instability at the failure initiation site prior to the failure initiation becomes possible. The arrival of the instability signal may result in some decrease of the stress difference in the vicinity of the failure site which, in turn, will result in the increase of the apparent failure threshold.

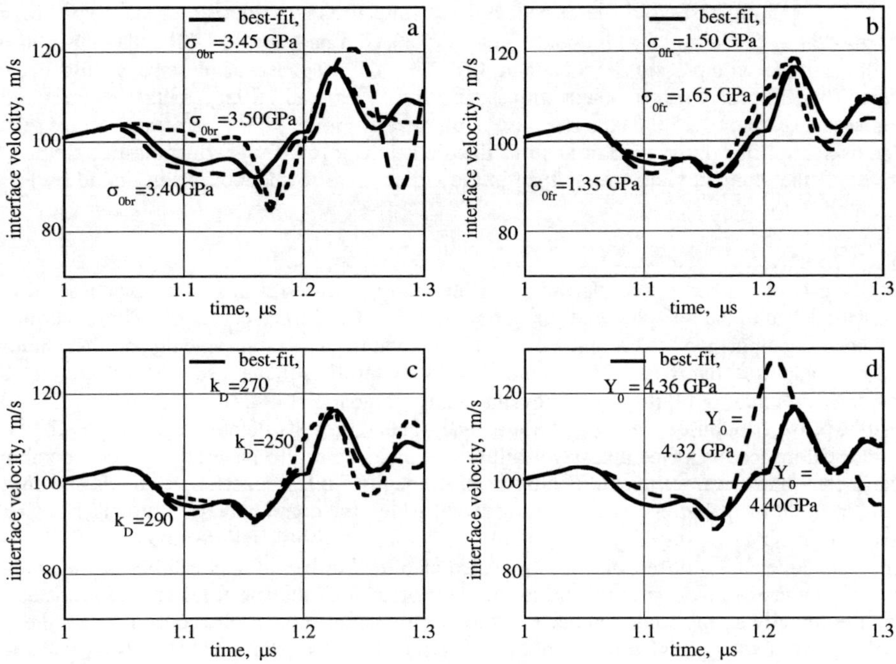

Figure 6. Computer simulation of the same, as in Fig. 5, impact. The influence of the variations of the failure parameters, σ_{0br} (a), σ_{0fr} (b), k_D (c), and yield stress Y_0, on the on the departure of the simulation from the best fit.

To summarize, both the alumina compressive failure threshold (failure initiation in compression) the trajectory of the comminuted alumina particles are described in the principal stress space by the linear equations with equal slopes

$$\sigma_{1br} = \sigma_{0br} + f \cdot \sigma_2 = 3.45 + 1.86\sigma_2 \text{ [GPa]}, \tag{8}$$

$$\sigma_{1fr} = \sigma_{0fr} + f \cdot \sigma_2 = 1.3 + 1.86\sigma_2 \text{ [GPa]}, \tag{9}$$

The ratio between alumina comminuting (cracking) level D and inelastic strain ε_{ie} is found equal to

$$\frac{dD}{d\varepsilon_{ie}} = k_D = 270 \pm 20. \qquad (10)$$

No correlation between the sample density (related to the presence of flaws micro-cracks in the sample) and the failure kinetics parameter k_D chosen for failure description of the sample was found.

The failure threshold described by (8) intersects the ductile yield (von Mises) line, σ_1=4.3+σ_2 at the point with coordinates σ_1=5.33 GPa and σ_2=1.01 GPa. The latter means that under radial compression higher then 1.01 GPa the response of alumina should became ductile. This is in some disagreement with static data of Heard and Cline [5] found that up to radial compression of about 1.25 GPa the response of alumina is brittle. The disagreement seems related to the oversimplified description of alumina ductile behavior (elastic - perfect plastic). Some 5-% increase of the alumina yield strength shifts the radial stress towards the value found by Heard and Cline [5].

CONCLUSIONS

We have studied the brittle failure of alumina in the series of impact experiments with plane-parallel alumina samples and copper impactors of different curvature. The experiments have been accompanied with continuous VISAR monitoring of the velocity of the alumina-window (sapphire) interface. The recorded velocity profiles have been reproduced in 2-D numerical simulations of the impact events using commercial AUTODYN code with added constitutive description of the brittle failure. The obtained results demonstrate the possibility of accurate determination of compressive failure threshold of brittle materials by such combined experimental/numerical technique. The method, although with lesser accuracy, allows one both to locate, in the principal stress space, the surface at which the comminuted brittle material finally arrives and to obtain an important information concerning the brittle failure kinetics.

Implementation of the work has revealed also the borders of applicability of the method. The main method limitation is related to the development of the surface corrugation instability when the velocity of the circumference of the impactor-sample contact area becomes less then the sample speed of sound (higher than that of impactor). Due to this limitation the stress states with radial stress σ_2 close to zero cannot be tested: approaching the σ_1 axis requires decreasing both the impact velocity V and the impactor radius R whose product, however, should not be much smaller than RV=74.

Increasing the impactor radius does not meet principal difficulties: together with the increase of both the sample and the window thickness it will result in the longer time internal suitable for failure observation. The increase of the impactor radius, however, will require producing the impactors with curved surfaces of optic quality. The latter is hardly possible with standard CNC turning machines.

Even with these limitations the suggested experimental/numerical technique is a useful tool for study of constitutive behavior of brittle solids capable of improvement by tightening the control of the impactor shape [16] and by detail description of the brittle failure process.

ACKNOWLEDGMENT

The work is partially supported by the Basic Research Found of the Israeli Ministry of Defense within the framework of Project No. 84297-301.

REFERENCES

[1]Z. Rosenberg, On the relation between the Hugoniot Elastic Limit and the yield strength of brittle materials, *J. Appl. Phys.*, 74 752 (1993).

[2]Z. Rosenberg, On the shear strength of shock loaded brittle solids, *J. Appl. Phys.*, 76 1543 (1994).

[3] E. B Zaretsky, V.E. Paris, G.I. Kanel, and A.S. Savinykh, Evidence of ductile and brittle response of ceramics under shock wave loading, *J. Phys. IV France*, 110, 917-922 (2003).

[4] J.U. Cazamias, W.D. Reinhart, C.H. Konrad, L.C. Chhabildas, and Bless S.J., Bar impact tests on alumina (AD995), in *Shock Compression of Condensed Matter – 2001,* edited by M. D. Furnish, N. N. Thadhani, and Y. Horie, American Institute of Physics 2002, 787-790.

[5] H. C. Heard and C.F. Cline, *J. Appl. Phys.* Vol 74, pp 752(1980.

[6] E. B. Zaretsky and G.I. Kanel, Evidence of ductile response of alumina ceramic under shock wave compression, *Applied physics letters*, Vol 81,7(2002).

[7] C.H.M. Simha, *High Rate Loading of a High Purity Ceramic - ID Stress Experiments and Constitutive Modeling,* Ph.D. Thesis, U.T. Austin, 1998.

[8] L. M. Barker and R.E. Hollenbach, Laser interferometer for measuring high velocities of any reflecting surface. *J. Appl. Phys.*, **43**, 4669. (1972).

[9]AUTODYN: interactive Non-Linear Dynamic Analysis Software, Ver 5.0.01, Century Dynamic Inc., 2005 (http://www.centdyn.com)

[10]D.E. Grady, Experimental analysis of spherical wave propagation, Journal of Geophysical Research, 78, N8, 1299-1307 (1973).

[11]S.K. Godunov, A.A. Deribas, A.V. Zabrodin, and N.S. Kozin, Hydrodynamic effects in colliding solids, *Journ. Comp. Phys*, 5 (N3), 517-539, 1970.

[12]R.E. Setchell, Refractive index of sapphire at 532 nm under shock compression and release *J. Appl. Phys.*, **91**, 2833-41. (2002).

[13]J. Wackerle, H.L. Stacy, and J.C. Dallman, Refractive index effects for shocked windows in interface velocimetry, SPIE Vol. 832, *High Speed Photography, Videography, and Photonics V-1987*, 72-82.

[14]M.F. Ashby and C.G. Sammis, The damage mechanics of brittle solids in compression, *PAGEOPH*, **133** (3), 490-521 (1990).

[15]M.L. Wilkins. Computer simulation of dynamic phenomena, *Springer*, Berlin, 1999.

[16]D.M. Anthony. Engineering metrology, *Pergamon Press*, Oxford, New-York, Frankfurt, Toronto, 1986.

DYNAMIC RESPONSE OF B_4C-SiC CERAMIC COMPOSITES

S. Hayun, N. Frage, M. P. Dariel,
Department of Materials Engineering, Ben-Gurion University of the Negev, P. 0. Box 653, Beer-Sheva 84105, Israel,

E. Zaretsky
Department of Mechanical Engineering, Ben-Gurion University of the Negev, P. 0. Box 653, Beer-Sheva 84105, Israel,

Y. Ashuah
Armament Development Authority, P.O. Box 2250, Haifa 32100, Israel

ABSTRACT

The dynamic response of boron carbide-silicon carbide (B_4C-SiC) composites was studied in planar impact experiments, using a 25-mm gas gun accelerating 1.5-mm thick tungsten impactors to velocities in the 750-850 m/s range. The velocity of composite sample-PMMA window interface was continuously monitored by Velocity Interferometer System for Any Reflector (VISAR). The composites were prepared by pressureless infiltration of porous B_4C ceramic preforms by molten silicon with and without carbon addition. During the infiltration process, molten silicon reacts with the B_4C preform and with free carbon, resulting in the formation of SiC. The Hugoniot Elastic Limit values, σ_{HEL}, of the ceramic samples were derived from the records of the velocity of the sample-window interface. The σ_{HEL} values were in the 11-15 GPa range and are affected by the presence of the residual silicon and by the morphology of the newly formed SiC. Analysis of the inelastic part of the VISAR records allowed concluding that the presence of residual silicon controls the failure of the B_4C-SiC composite in compression.

INTRODUCTION

Hot pressed boron carbide (B_4C) has excellent dynamic properties. Its low, about 2.50 g/cm^3, density and high Hugoniot Elastic Limit (HEL), 17-20 GPa [1-4], provide a firm basis for its use in light armor. Higher than 2200°C sintering temperature of the boron carbide makes the fabrication of high performance (low porosity) boron carbide ceramics a complicated and expensive process. This provides motivation for producing dense boron carbide-based ceramics by methods simpler than sintering under pressure. Reaction bonding of silicon carbide (RBSC) allows obtaining fully dense composites [5-7] at a cost much lower than for hot pressing. In this process, a porous green SiC-carbon body undergoes pressureless infiltration by molten silicon. It has also been reported [8, 9] that B_4C preforms may be used for fabricating reaction bonded composites. In this case, the microstructure of the composite consists of the initial boron carbide particles, the newly formed SiC, and residual silicon. Reaction bonded boron carbide may be produced by infiltrating B_4C preforms with molten Si in the absence free carbon. Molten Si reacts directly with carbon present in the carbide leading to the formation of SiC. The presence of the residual silicon pockets along the boundaries of SiC grains is the inevitable outcome of such process. These low-strength metal inclusions are the sources of the early activation of inelastic deformation in the composite. In order to decrease the amount of residual silicon in the final product, some free carbon may be added to the initial boron carbide powder. In that case two carbon sources operate in the composite and one may expect different morphologies for the growing SiC particles, leading to differences in the dynamic response of the composites with and without excess carbon. In the present work, the influence of the SiC amount and morphology on the evolution of the dynamic response of two kinds of boron carbide-silicon carbide composites was studied by planar impact experiments. The two kinds of composites show a remarkable difference in their dynamic behavior.

MATERIALS AND PLANAR IMPACT EXPERIMENTS

Fabrication of composites.

The two groups of composites were produced by the same technique from two sources of raw material. The first kind was prepared from a relatively coarse, 5μm mean size, particles with a narrow particle distribution. The boron carbide content in the powder supplied by Mudanjiang Abrasives And Grinding Tools Co., Mudanjiang, People Republic of China was 95 mass %. The composites produced from this powder are denoted as CB. Preforms of the second group, labeled as SB, were prepared from a 97-% purity grade HS finer, maximum 3-μm size particles that had an extended "tail" towards 0.2-μm size, powder supplied by H.C.Starck, Kelsterbach, Germany. The latter material is used for sintering applications while the former is for grinding. Compacts of 20 mm diameter and of 3 mm high were obtained from both B_4C powders by uniaxial compaction (10 MPa). The compacts were sintered at 2173-2373 K for 30 min in order to yield preforms with 20, 30, or-40 vol.% porosity. For fabrication of the carbon-containing composites, CBC and SBC, respectively, the free carbon was added to the porous preforms by their saturation with a water-sugar solution (50:50). The preforms saturated with sugar solution were dried and heat treated at 773K in order to complete the pyrolisis of the sugar. The preforms with and without excess carbon were subsequently infiltrated with liquid Si in a vacuum furnace (10^{-4} torr) at 1723K for 20 min.

Composites microstructure

The SEM (JEOL-35) study of the microstructure of the samples was done in parallel with image analysis performed by Thixomet software (developed at the St.-Petersburg State Polytechnic University, Department of Steel and Alloys), in order to determine the amount of SiC and residual Si in the composites (listed in Tab. 1). Density measurements were used for verification of the image analysis data.

Figure 1 illustrates the difference between the CB and SB composites and the changes induced by the presence of the carbon excess. It is apparent that in spite of the relatively small difference in the mean grain size of the original powder used for the CB and SB samples, the

Table 1. Average phase distribution in carbon-free and carbon-containing composites

Samples	Free C, vol.%	B_4C, vol.%	Si, vol.%	SiC, vol. %	Samples	Free C, vol.%	B_4C, vol.%	Si, vol.%	SiC, vol.%
SB	0	80±3	7±1	13±1	CB	0	80±3	12±1	8±1
	0	70±3	13±1	17±1		0	70±3	20±1	10±1
	0	60±3	20±1	20±1		0	60±3	28±1	12±1
SBC	3±0.5	80±3	8±1	12±1	CBC	3±0.5	80±3	12±1	8±1
	3±0.5	70±3	13±1	17±1		4±0.5	70±3	21±1	9±1
	5±0.5	60±3	16±1	24±1		6±0.5	60±3	30±1	10±1

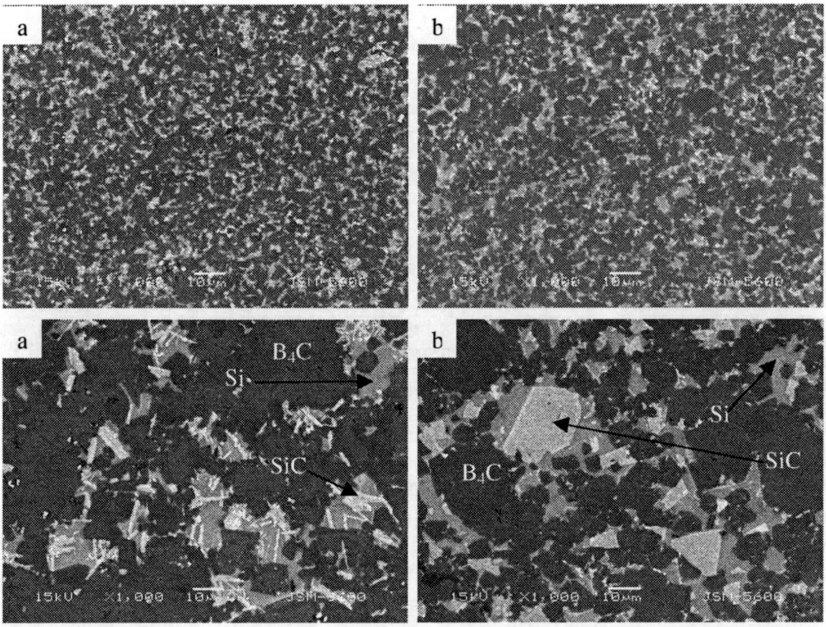

Figure 1. Microstructure (×1000) of two types of B4C-Si composite based on boron carbide preforms having equal (20%) porosity but produced from different, CB (bottom) and SB, boron carbide powder without carbon addition (a) and at the presence of free carbon (b). For the case of SB material the magnification of about ×5000 is required in order to observe the phase pattern similar to that of CB sample.

pores in the SB material are smaller and are distributed in a more homogeneous manner. The presence of the excess carbon assists the conversion of the free silicon into SiC , thereby decreasing the amount of the residual silicon. It also changes the morphology of the SiC precipitates: instead of SiC needles starting from B_4C-Si boundary (the only source of carbon supply in the absence of free carbon) and reinforcing the "soft" Si inclusion, the presence of free carbon favors the growth of SiC in the Si pockets, resulting in the formation of the globular SiC particulate along or within the original Si regions.

Physical properties of the composites

The longitudinal C_l and shear C_s speeds of sound in the studied materials were determined by the pulse-echo technique using 5 MHz probe. The Young E and shear G moduli, were derived (Table 2) from these data using density values measured by the Archimedes method.

The elastic moduli of the composites display a modest (at the border of the error of the moduli determination) shift towards values lower than those expected from the rule of mixtures. The shift is attributed to the residual porosity (estimated to be less than 1% for all the samples) of the composites.

Table 2. Physical properties of studied composites

Materials*	Density, g/cm^3	C_l, km/s	C_s, km/s	Poisson's ratio	G, GPa	E, GPa
CB20	2.557±0.014	13.11±0.27	8.22±0.13	0.17±0.01	173±5	407±15
CB30	2.555±0.027	12.45±0.28	7.67±0.12	0.19±0.02	150±6	359±15
CB40	2.547±0.005	11.94±0.09	7.36±0.07	0.19±0.01	138±2	330± 5
CB20C	2.565±0.002	12.99±0.08	8.16±0.14	0.17±0.02	171±6	401± 6
CB30C	2.618±0008	12.12±0.10	7.50±0.07	0.19±0.01	148±3	351± 6
CB40C	2.552±0.022	12.43±0.21	7.64±0.12	0.20±0.02	149±6	356±11
SB20	2.582±0.060	13.24±0.05	8.21±0.04	0.19±0.01	174±3	413± 9
SB30	2.606±0.012	12.52±0.09	7.75±0.04	0.19±0.01	157±2	372± 4
SB40	2.583±0.001	12.55±0.04	7.50±0.06	0.21±0.03	147±4	358± 4
SB20C	2.601±0.007	13.18±0.06	8.14±0.02	0.19±0.01	173±6	411± 6
SB30C	2.627±0008	13.45±0.10	7.96±0.07	0.23±0.01	167 ±3	400± 6

*Arabic numerals correspond to the average porosity of the B_4C preform.; the addition of C is marked by corresponding Latin letter.

Planar impact experiments and ballistic tests

The dynamic response of the boron carbide-silicon composites was studied in the planar impact experiments with 25-mm, 3.5-m length gas gun. The gun is capable of accelerating hollow PMMA projectile equipped with 1.5 mm thick, 24.5 mm diameter impactor to a velocity of about 900m/s. The impactors were made of a tungsten heavy alloy (WHA) having dynamic mechanical properties close to those attributed to WHA by Steinberg et al. [11]. The 20-mm diameter composite samples prepared from boron carbide preforms with approximately 20, 30, and 40 vol.% initial porosity, with and without the addition of carbon, infiltrated with molten silicon were ground to about 3mm thickness and polished to ensure plane-parallel surfaces within 0.005 mm. The 6-mm thickness PMMA windows, with stuck 7-μm aluminum foil as light reflector, were glued to the rear (un-impacted) sample surface, in order to prevent the loss of the surface reflectivity at the arrival of the inelastic wave at the sample free surface. In the single shot at a 3-mm, 99.999-% purity silicon sample (25-mm silicon rod, Alpha Aesar Company) a 5-mm sapphire window was used. The velocities of the impacted sample-window interface were continuously monitored by VISAR [12] having interferometer constant 392.7 m/s per fringe. The sample holder was equipped with electrical pins whose sequential shortening allowed measuring the impact velocity and triggering the Tectronix 7104 1-MHz oscilloscope. The impact velocity in the experiments was about 750–850 m/s. The impactor/sample misalignment did not exceed 1 mrad in all experiments.

In order to evaluate the ballistic weight efficiency of the both CB and SB ceramics with respect to other armor materials, the "thick backing" testing configuration suggested by Rosenberg et al.[13] was used. The tiles were bonded by a cyanoacrylate glue to a semi-infinite aluminum 2024 T351 block and shot with 7.62×54 AP projectiles having velocities ranged between 790-800 m/s. The depth-of-penetration P of the projectile in the aluminum backing was measured. The ballistic weight efficiency η of a tile was calculated using the expression

$$\eta = \frac{(P_0 - P)\cdot \rho_{Al}}{t_{cer}\cdot \rho_{cer}},$$

where P_0 is the reference depth-of-penetration in the semi-infinite backing at the absence of a tile, t_{cer} is the thickness of the tile, and ρ_{Al} ,and ρ_{cer} are the densities of the backing and the tile, respectively.

EXPERIMENTAL RESULTS AND DISCUSSION

The sample-window interface velocity profiles obtained by the planar impact experiments with B_4C-Si samples are shown in Fig. 2. The reproducibility of the profiles features was checked in repeated shots for several compositions and was found excellent. All the velocity profiles of Fig. 2 have a similar structure, namely an elastic jump of some 700-750 m/s, followed by a velocity ramp, which, in turn, is followed by a more (SB ceramics) or less (CB materials) pronounced velocity decrease. This first maximum stress may be interpreted as Hugoniot elastic Limit of the ceramics. The corresponding stress σ_{HEL} values close, for SB20 and SB20C materials, to the HEL values of pure dense boron carbide [2, 3, 14], are listed in the Tab. 3. The minor, 10-20 m/s, velocity spike coinciding with the elastic jump is attributed to the presence in the composite of a small amount of residual silicon whose inelastic deformation starts with a similar spike (the velocity profile at the center of the bottom part of Fig. 2). The second, and the larger, velocity ramp, related to the inelastic deformation increase, is separated from the elastic one by a time interval much longer than expected from the difference between the times of the propagation of the elastic ($t_e = \delta/C_l$) and the bulk ($t_{pl} = \delta/C_0$) waves through the sample of the thickness δ. The protracted release parts of the profiles with no sign of the elastic unloading allows one to assume that the regions of the samples adjacent to the impactors were seriously damaged, possibly comminuted, under compression. Such assumption requires to relate the inelastic velocity ramp to reverberation of elastic signal between the sample-window interface and the damage front as it is shown in Fig. 3 for two velocity profiles typical for each, CB and SB, group of the composites. Estimate of the damage front velocity C_D from the timing of inelastic ramp (Fig. 3) yields $C_D \approx 7$ km/s and $C_D \approx 5$ km/s for CB and SB groups of ceramics, respectively. Such difference between the damage front velocities corresponds to the difference between damage rates in the two types of materials: the higher damage rate in SB ceramics is accompanied by faster stress relaxation and, thus, impedes the damage wave propagation. The velocity drop (stronger in SB materials) preceding the inelastic velocity ramp corresponds to the release signals accompanying the material damage, while the initial slope of the velocity decrease is related to the kinetics of the damage.

The velocity records with aid of the simple wave approximation [15] allows studying the influence of the structure and composition of the ceramic samples on their strength in compression.

Table 3. Hugonio Elastic Limits of the studied ceramics

Material average. porosity, %	CB	CBC	SB	SBC
	σ_{HEL}, GPa			
20	13.5±0.3	12.1±0.4	13.5±0.3	14.8±0.5
30	12.1±0.4	10.5±0.5	11.5±0.5	14.6±0.6
40	10.1±0.6	10.1±0.5	13.0±0.4	13.9±0.3

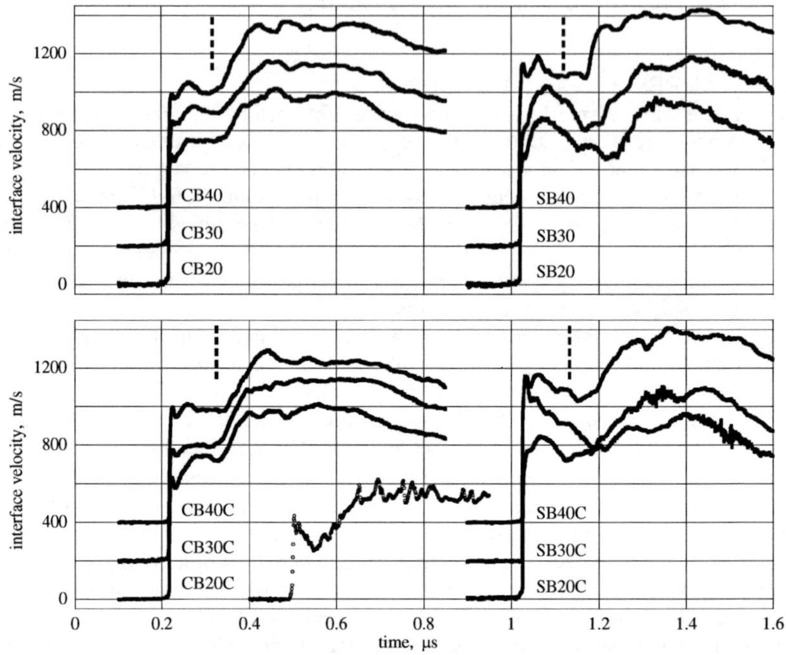

Figure 2. Experimental velocity profiles obtained from the interface between ceramic sample and PMMA window after planar impacts produced by 1.5-mm WHA samples. Part of the velocity profiles are shifted up for clarity. The tested material is marked at the bottom of each profile. Dashed line shows time of arrival of an inelastic wave estimated on the base of the bulk speed of sound, $C_0=(C_l^2-4/3C_s^2)^{1/2}$, for undamaged material. At the central part of the bottom plot the impact response of pure silicon characterized by the spike at the elastic wave front is shown.

Since the approximation relates the recorded VISAR signal to the centered wave generated at the impacted sample surface, the validity of the approximation is limited by the first maximum of the recorded velocity. At this instant, the unloading signals arriving from the damaged sample region start to contribute to the recorded signal. Part of the signal prior to the first velocity maximum provides important information regarding the material state preceding the initiation of the damage. The compressive part of the loading pulse immediately behind the elastic jump may be considered as a simple centered wave described by a fan of characteristics. For simple waves, the longitudinal stress increments $d\sigma_x$, the strain increments $d\varepsilon_x=dV/V_0$, and the pressure are related by the equations

$$d\varepsilon_x = -\frac{du_p}{a_\sigma}, \tag{1a}$$

$$d\sigma_x = \rho_0 a_\sigma^2 d\varepsilon_x, \tag{1b}$$

$$p = \rho_0 c_b^2 \frac{\varepsilon_x}{(1 - s\varepsilon_x)^2}, \tag{1c}$$

where a_σ is the Lagrangian velocity of propagation of the part of the compression wave at longitudinal stress σ_x [15] corresponding to the particle velocity u_p, and s is the slope of the linear Hugoniot $D = C_0 + Su$ which can be considered as equal for all the studied composites because their similar composition.

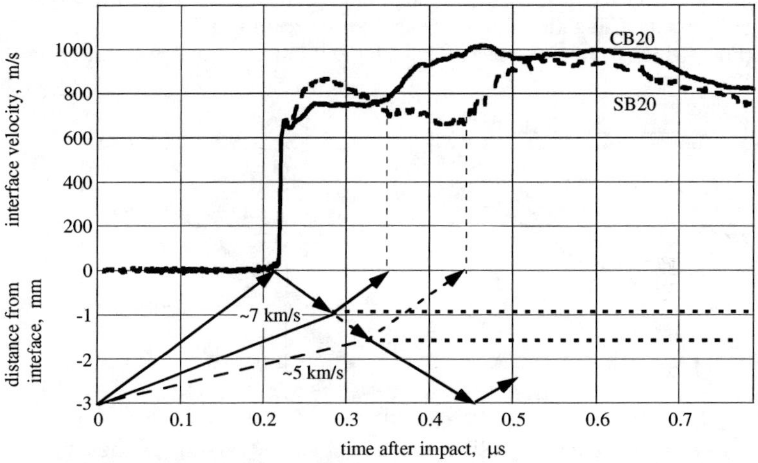

Figure 3. Interface velocity profiles (upper part of the graph) obtained in planar impact experiments with CB20 and SB20 ceramics. The bottom part shows the corresponding distance h (Lagrangian coordinate)-time t diagrams. After reflecting, as an unloading wave, from the sample-window interface ($h = 0$) a part of the elastic wave front is reflected from the interface between damaged (below the corresponding dotted line) and undamaged material. The arrival of the elastic unloading stops the sample damaging. The reflected part of the wave returns to the sample-window interface bringing the information about the damage front. The transmitted part of the wave, after reflection from the impactor-sample interface ($h = -3$ mm), returns, as release wave, to the sample window interface.

The stress σ_x is assumed to be positive under compression. For a centered simple wave, the propagation velocity a_σ is

$$a_\sigma = C_l \frac{1}{1 + C_l t_\sigma / h}, \tag{2}$$

where h is the Lagrangian coordinate of the point at which the stress history $\sigma_x(t)$ of the compressive wave is analyzed, and t_σ is the time interval after the elastic precursor wave front.

Calculating the particle velocity u_p from recorded velocity profiles with aid of known PMMA Hugoniot [16] and integrating Eq.(1) yields a parametric form of the deviator stress-strain dependence:

$$\varepsilon_x = \varepsilon_x(t), \tag{3a}$$

$$s_x = s_x(t) = \sigma_x(t) - p(t). \tag{3b}$$

Figure 4 shows the deviator-strain diagrams obtained for CB and SB ceramics of 20 and 30-% initial porosity using the simple wave approximation. It is interesting to note that SB ceramics (Fig. 4, left) having higher HEL display the failure initiation at lower values of stress deviator.

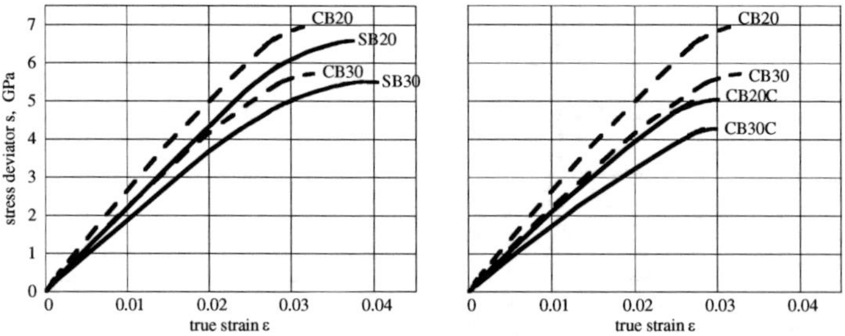

Figure 4. Compressive curves of CB and SB ceramics (at the absence of carbon addition) obtained from recorded velocity profiles with aid of simple wave approximation (left), and the influence of carbon addition on the strength of CB ceramics (right).

The maximum strain the ceramics may acquire prior to the failure is, however, higher in SB ceramics in which the amount of residual silicon (see Tab. 1) is 1.4-1.9 times lower than in CB materials. This confirms Grady's suggestion [3], that the residual metallic phase bears the responsibility for failure initiation in hot-pressed ceramics. Adding free carbon (Fig. 4, right) results in the decrease of both the maximum stress deviator and the strain-to-failure. The values of inelastic strain to failure, determined from the compressive curves (similar to those of Fig. 4) of the studied ceramics, differs for SB20 and SB30 (both without carbon additions) and the rest of the composites. The maximum inelastic deformation preceding failure reached in these two materials is close to 1%. In the other samples it does not exceed 0.4%. This means that in the SB20 and SB30 ceramics the initiation of failure (cracking, comminuting) under conditions of the fast impact-induced overstressing is delayed because of the absence of the sites suitable for the crack nucleation in the more homogeneous microstructure. When crack nucleation conditions are achieved in a sample cross-section (the loading is planar), they occur in a multitude of the cross-section points leading to the fast (faster than in the rest of the samples) comminuting of the sample.

In spite of distinctive differences in the dynamic strength, HELs and failure kinetics, both the CB and SB, of ceramics display a similar ballistic efficiency, independent of the presence or absence of free carbon. The ballistic weight efficiency of the CB 5-mm ceramic tiles with 20 to

30-% initial porosity and with carbon content that varied from zero to 8%, as well as that of SB ceramics with 30-% initial porosity was independent of both material type and composition and equal to $\eta=7\pm0.5$ (good armor alumina tiles yield efficiencies around 5.5, while boron carbide can reach $\eta=8$). This ballistic efficiency for ceramic tiles, prepared from low-cost (CB, CBC) raw materials and by sintering at relatively low temperature, may be considered as a promising result.

CONCLUSIVE REMARKS

The present study of boron carbide-silicon carbide composites shows that the study of the dynamic mechanical properties of these new materials by planar impact experiments with VISAR diagnostics provides useful links between their microstructure and the features of their high-strain rate behavior. Moreover, the velocity profiles yield a firm basis for the search of proper constitutive description of these materials which, in turn, may allow designing the armor tiles with improved ballistic efficiency.

ACKNOWLEDGMENT

This work was partially supported by the Basic Research Found of the Israeli Ministry of Defense within the framework of Project No. 84297-301.

REFERENCES

[1]W.H. Gust and E.B. Royce, *J. Appl. Phys.* 42, 276–295 (1971).

[2]M.E. Kipp and D.E. Grady, in: *Shock Compression of Condensed Matter—1989* (eds. S.C.Schmidt, J.N. Johnson, and L.W. Davison) North-Holland, Amsterdam, pp. 377-380 (1990).

[3]D.E. Grady, in: *High-Pressure Science and Technology—1993* (eds. S.C. Schmidt, J. W. Shaner, G.A. Samara, and M. Ross) American Institute of Physics, New York, pp. 741–744 (1994)

[4]N.S. Brar, Z. Rosenberg, and S.J. Bless, in: *Shock Compression of Condensed Matter—1991* (eds. S.C. Schmidt, R.D. Dick, J.W. Forbes, and D.G. Tasker), Elsevier Science Publishers, B.V., North-Holland, Amsterdam, pp. 467-470 (1992).

[5]K.M. Taylor, "Cold Molded Dense Silicon Carbide Articles and methods of making the same." U.S. Pat. No. 3 205 043, Sept. 7 1965.

[6]P.P. Popper, "Production of Dense Bodies of Silicon Carbide. " U.S. Pat. No. 3 275 722 Sept. 27, 1966.

[7]C.W. Forrest, "Manufacture of Dense Bodies of Silicon Carbide. " U.S. Pat. No. 3 495 939 Feb. 17. 1970.

[8]M.K. Aghajanian, B.N.Morgan, J.R.Singh, A new family of reaction bonded ceramics for armor applications, in "*Ceramic Armor by Design*",Eds. J.W. McCauley et. al, Ceramic Transactions, Vol. 134, American Ceramic Society, 2002, pp.527-540.

[9]M.K. Aghajanian, A.J. McCormick, B.N.Morgan, A.F. Liszkiewicz Jr., "Boron carbide composite bodies, and methods for making same", U.S.Patent Application 00400065868, (2004)

[10]D.E. Grady, shock profile studies on selected silicon carbide ceramics with application to dynamic yield mechanism, in *Shock Compression of Condensed Matter – 1999* (eds. M. D. Furnish, L. C. Chhabildas, and R. S. Hixson) American Institute of Physics 2000, 629-632.

[11]Steinberg D.J., Cochran S.J. and Guinan M.W., A constitutive model for metals applicable at high-strain rate, *J. Appl. Phys.*, 51, 1498-1504 (1980).

[12]L.M. Barker, and R.E. Hollenbach, Laser interferometer for meshuring high velocities of any reflecting surface, *J.Appl.Phys.*, 43, 4669-4675 (1972).

[13]Z. Rosenberg, S.J. Bless, Y. Yeshurun and K. Okajima, A new definition of the ballistic efficiency of brittle materials based on the use of thick backing plates, Proc. of the "*Impact 87*" Conference in Bremen, FRG, 18-22 May 1987, eds. C.Y. Chiem, H.D. Kunze and L.W. Meyer, p. 491.

[14]T.J.Vogler, W.D. Reinhart, and L.C. Chhabildas, Dynamic behavior of boron carbide, *J.Appl.Phys*, 95 (2004) 4173-4183

[15]E.B. Zaretsky, G.I. Kanel, S.V. Razorenov, and K. Baumung, Impact strength properties of nickel-based refractory superalloys at normal and elevated temperatures, *International Journal of Impact Engineering*, 31 (2005) 41-54.

[16]L.M. Barker, and R.E. Hollenbach, Shock wave studies of PMMA, fused silica, and sapphire, *J.Appl.Phys.*, 41 (1970)4208-4226

Author Index

Anderson, A., ...63
Ashuah, Y., ...147

Brennan, R., ...103
Butt, D.P., ...91

Colombo, P., ...79
Crane, S., ...91
Cutler, R.A., ...63, 91

Dariel, M.P., ...37, 147
Demirbas, M.V., ...109
Dilman, H., ...37

Fisher, D., ...19
Flinders, R.M., ...63, 91
Frage, N., ...37, 147

Gahr, M., ...45

Haber, R., ...103
Haber, R.A., ...109
Hamling, J.A., ...55
Hartnett, T.M., ...19
Hayun, S., ...37, 147
Heidenreich, B., ...45
Hurley, J.P., ...55

Kanel, G.I., ...131
Kim, D.Y., ...123
Kim, J.H., ...123
Kim, Y.G., ...123

McCauley, J., ...103
Medvedovski, E., ...3, 45, 79

Niesz, D., ...103

Pabit, E., ...91
Paris, V.E., ...131

Rajendran, A., ...131
Ray, D., ...63, 91

Seiben, K., ...91
Singh, V., ...55
Sunne, W., ...19

Tourbabin, V., ...37

Warner, C.T., ...19

Zaretsky, E., ...147
Zaretsky, E.B., ...131
Zordan, F., ...79